Limit States of Compressed Concrete Elements by Transverse Deformations

Iakov Iskhakov
Ilya Frolov and **Yuri Ribakov**
Ariel University, Israel

CRC Press
Taylor & Francis Group
Boca Raton London New York

CRC Press is an imprint of the
Taylor & Francis Group, an **informa** business
A SCIENCE PUBLISHERS BOOK

First edition published 2025
by CRC Press
2385 NW Executive Center Drive, Suite 320, Boca Raton FL 33431

and by CRC Press
4 Park Square, Milton Park, Abingdon, Oxon, OX14 4RN

CRC Press is an imprint of Taylor & Francis Group, LLC

Library of Congress Cataloging-in-Publication Data (applied for)

ISBN: 978-1-032-73891-8 (hbk)
ISBN: 978-1-032-75531-1 (pbk)
ISBN: 978-1-003-47438-8 (ebk)

DOI: 10.1201/9781003474388

Typeset in Times New Roman
by Prime Publishing Services

Preface

In the modern theory of compressed concrete elements, the main attention is usually paid to longitudinal deformations, whereas the transverse ones are rarely considered just within the Poisson's coefficient limits (i.e., inelastic compressed concrete behavior in transverse direction). However, transverse deformations significantly develop beyond the limits of elastic deformations, corresponding to the Poisson's coefficient. If the deformations of compressed concrete are inelastic, the ratio between longitudinal and transverse deformations is obtained. In this case, the ratio increases non-linearly leading to initiation and development of longitudinal cracks in compressed elements. Experimental and numerical investigations of transverse deformations in the inelastic stage of these elements showed that it is necessary to consider the crack propagation, because it can lead to ultimate limit state by transverse deformations. This book presents a novel approach, based on simultaneous consideration of longitudinal and transverse deformations, as well as the appearance of cracks and their width and depth. It allows obtaining a complete compressed concrete elements' behavior pattern at all performance stages in two types of limit states (by longitudinal and transverse deformations). Consequently, a new ultimate limit state by the depth and width of cracks due to transverse deformations (additionally to that, by longitudinal ones) is proposed to be used in modern design practice.

Taking into account the fact that transverse tensile deformations in high-strength concrete are rather big, an additional experimental investigation was carried out to study the effectiveness of the above-mentioned ultimate limit state for behavior of this concrete in the inelastic stage. With this aim, a novel method, based on testing normal-strength concrete cylindrical specimens under compression at relatively high loading rates, is proposed for investigating compressed high-strength concrete elements. The experimental and numerical investigations were carried out to confirm the theoretical model of cracking and failure in compressed concrete elements, taking into account the ultimate limit states in both longitudinal and transverse directions.

The book also considers the fact that the design of compressed concrete elements can be carried out based on the theoretical stress-strain model, without empirical coefficients, that was previously proposed by Iskhakov. This model includes the following main parameters describing concrete behavior: stresses and strains, corresponding to the border between the elastic and inelastic behavior stages of a concrete specimen, ultimate elastic strains, and stress–strain behavior at the post-peak region. Particular attention is focused on the descending branch of the stress-strain diagram, as well as on the analysis of concrete elastic and plastic potentials. These potentials are important for investigating the dynamic response of the concrete element section, as well as for concrete creep analysis.

Contents

1

Introduction

Concrete is one of the most commonly used building materials in the world. It has very effective behavior under compression (in comparison with other loading types) and comparatively low production cost. Therefore, concrete is successfully used in structural elements subjected to different types of loadings, like compression, bending, etc.

Structural design of concrete and reinforced concrete (RC) elements is based on their limit states, presented in such modern codes as *Eurocode 2* and *SI 466 Part 1* [1, 2]. These limit states are defined just from the viewpoint of concrete performance in the longitudinal direction, i.e., the stress-longitudinal strain relationship for the element section zone under compression. This book presents a discussion of a new limit state for concrete structures by transverse deformations due to longitudinal compression loads. The main aim is to provide a complete view of the limit states, governing compressed concrete elements behavior.

The present research helps to minimize the number of empirical coefficients used in modern codes for the design of concrete elements. Since concrete is a heterogeneous non-isotropic stone-type material, it is difficult to find an accurate model describing its behavior. Consequently, known dependencies of concrete behavior include empirical coefficients proposed that are based on available experimental data. It leads to higher construction costs since the design approaches are not accurate.

An effective theoretical method, allowing to minimize the number of empirical coefficients in dependences used for design of concrete and RC structures was previously proposed by Iskhakov and Ribakov [3]. This method forms a basis for the new limit state proposed in this book.

2

Literature Review

2.1 Historical Overview

The problem of non-effective concrete behavior modeling was described in the beginning of the 20th century. A detailed overview of concrete and RC structures theory development up to the 1950s was presented by Hognestad [4]. Further, a summary of the overview is provided.

The concrete theory development started from the investigation of bent RC slabs [5, 6, 7, 8, 9, 10, 11, 12, 13]. This yielded to a "standard" compressed section theory, considering only the elastic stage of concrete performance (Figure 1) [7]. The mathematical interpretation of this theory is:

$$P = f_c A_c [1 + (n - 1)\rho_{st}] \quad ...(1)$$

where P is a load-bearing capacity;

f_c is the concrete compressive strength;

A_c is the concrete section area;

ρ_{st} is a ratio between the areas of the effective longitudinal reinforcement and the entire concrete section;

n is a modular ratio, $n = \frac{E_s}{E_c}$;

E_c is modulus of elasticity of concrete;

E_s is modulus of elasticity of reinforcement steel.

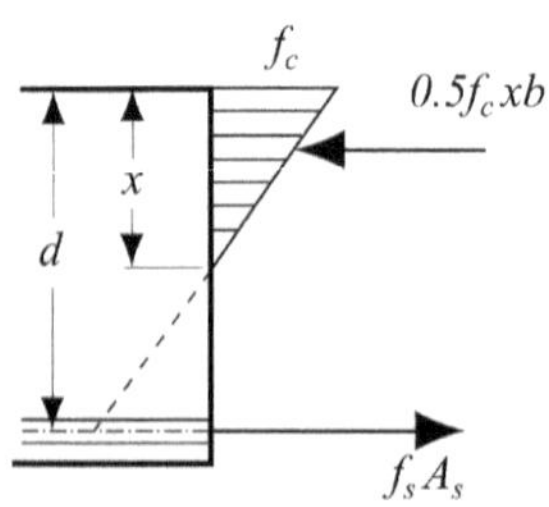

Fig. 1 Bent RC section stress diagram ("standard theory"), following [7].

The obvious problem of this approach was the overestimated safety factor, which was approximately close to 3. Extensive laboratory concrete studies had been carried out by different researchers (Emperger [11], Talbot [14, 15], Bach and Graf [16] etc.), in order to develop the proposed "standard" model.

Another problem of the "standard" design approach, caused by the lack of theoretical data in the initial period of concrete research, was neglecting the additional bending effect on stress distribution in RC columns design [4]. This important parameter was compensated just by the high safety factor. However, developing the Slope-Deflection Method [17] and the Moment Distribution Method [18, 19] in the second and fourth decades of the 20th century, correspondingly, partially solved this problem (the moments of inertia of an uncracked section were used). Further extensive tests of structures [20, 21] have shown that an application of Neumann's theory (linear stress distribution) [6] is satisfactory for structural analysis of indeterminate reinforced concrete structures if proper values of the modular ratio, *n*, are chosen.

Looking ahead, it should be mentioned that during the period from the 1920s to the 1950s, Slater, Lyse, Richart, and Staehle carried out many of experimental studies on axially loaded RC columns, continuing the earlier investigation conducted by McMillan [22], aimed at examining inelastic RC behavior. As a result, the following equation, including a higher safety factor for concrete than for reinforcement steel was obtained:

$$P = 0.225 f_{ck} A_c + 0.4 f_y A_s \qquad ...(2)$$

where f_{ck} is a characteristic compressive cylinder strength of concrete at 28 days;

f_y is a yield strength of longitudinal reinforcement.

This approach was later included into the ACI Building Code [23]. However, it was applicable just for inelastic design of axially loaded compressed RC structures (or elements with small eccentricities) and in the case of large eccentricities the "standard theory" was still used.

Further development of inelastic structural design included the following failure models and approaches:

- section tensile zone failure caused by the tensile steel yielding and corresponding change in the neutral axis position, leading to ultimate compressive strain in concrete;
- section compression zone failure caused by development of ultimate compressive strain in concrete before the tensile reinforcement steel reaches yielding;

- simultaneous section failure caused by reaching yielding in both tensile reinforcement steel and compressed concrete;
- compression failure of concrete with tensile reinforcement steel stresses higher than the yield point (under-reinforced section) [24];
- brittle failure caused by rupture of the tensile reinforcement steel immediately after the formation of cracks in the tensile zone of the concrete section (over-reinforced section) [24].

In a case of beams and slabs' section failure by tensile reinforcement, a rectangular stress-block for the concrete compressed zone was proposed [25] (Figure 2).

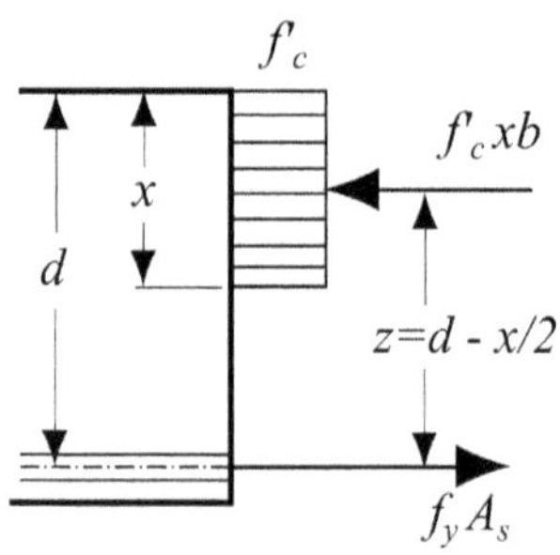

Fig. 2 Stress diagram of a bent RC section, following [25].

The expression of the moment for the cross-section without compressed reinforcement, that is still in use in modern design methods [2], was obtained from the following equilibrium equations:

$$\frac{x}{d} = 1 - \sqrt{1 - \frac{2M}{bd^2 f_c'}} \qquad ...(3)$$

$$f_s = \frac{M}{A_s z} \qquad ...(4)$$

where x is the concrete rectangular stress-block depth;
M is the bending moment;
f'_c is concrete compressive strength.
or in another form:

$$\frac{M}{bd^2} = \rho f_y \left(1 - \frac{1}{2}\rho \frac{f_y}{f_c'}\right) \qquad ...(5)$$

where $\rho = \dfrac{A_s}{bd}$.

Later this theory was experimentally verified by testing 10 continuous RC beams with different reinforcement types [26]. The section compressive strength was assumed equal to the prism strength, f_{ck}^{prism}, and the tensile stress to the reinforcement yield strength, f_y.

Mensch proposed a theory for the rectangular beam section for the tensile failure as well as for the compression one [27]. According to this theory, in case of the section tensile zone failure, ultimate moments are obtained based on the stress-block in the form of Ritter's second-order parabola [28]; however, without considering ultimate strains and Bernoulli's hypothesis (Figure 3a). In the case of section compression zone failure, it was assumed that the neutral axis reached the centroid of the tensile steel (Figure 3b). However, this limitation was too extreme, and it was weakened by Mensch by an empirical coefficient [27]:

$$\frac{M}{bd^2} = \frac{f_c'}{2.6} \quad \text{...(6)}$$

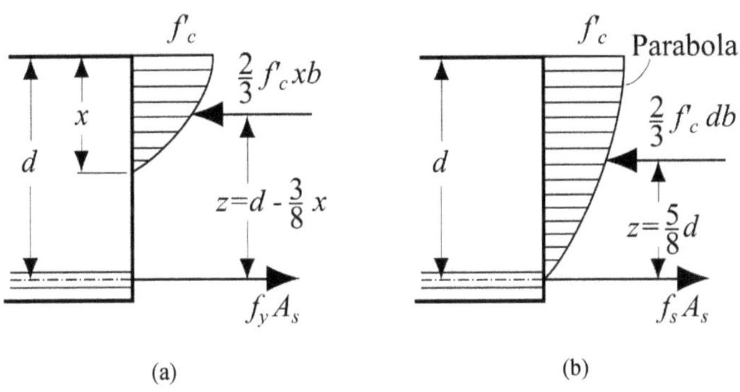

Fig. 3 Stress diagrams of a bent RC section, following [27]: (a) section tensile zone failure; (b) section compression zone failure.

In 1922, Kempton Dyson published an approach similar to the Mensch approach for calculating ultimate moments based on an elliptical stress-block [29] (Figures 4a and 4b). The equation for moments in the section compression zone failure case is:

$$\frac{M}{bd^2} = \frac{f_c'}{2.21} \quad \text{...(7)}$$

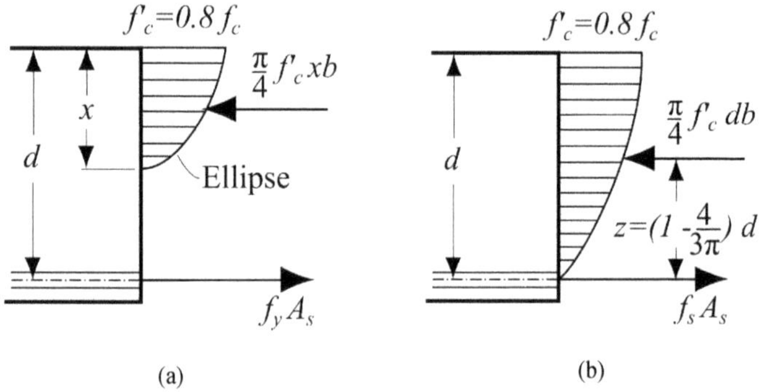

Fig. 4 Stress diagrams of a bent RC section, following [29]:
(a) section tensile zone failure; (b) section compression zone failure.

The ultimate theory developed by Stüssi [30] considered the ultimate moment in beams with the rectangular section and tensile reinforcement only. This method was widely used in later investigations. According to this theory an arbitrary form of the concrete stress-block for rectangular sections was characterized by constants k_1 and k_2, compressive strength, f'_c, and ultimate strain, $\varepsilon_{c\,ul}$ (Figure 5).

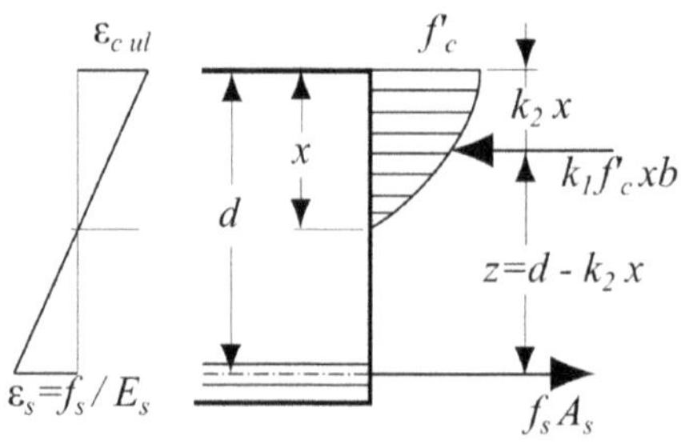

Fig. 5 Stress and strain diagrams of a bent RC section, following [30].

The main feature of this study was obtaining all of the above-mentioned parameters from testing compressed prisms. Thus, the parameters used by Stüssi were k_1 = 0.7…0.77, k_2 = 0.39…0.41, $\varepsilon_{c\,ul}$ = 2.0…2.5‰. Additionally, it was assumed that the compressive strength in flexure is equal to the prism strength, f_{ck}^{prism}. The following equilibrium equations were obtained:

$$A_s f_s = k_1\, bx f_{ck}^{prism} \qquad ...(8)$$

$$M = k_1\, bx f_{ck}^{prism}\,(d - k_2\, x) \qquad ...(9)$$

For the section tensile zone failure, the ultimate moment equation was found very similar to (5), especially considering Stüssi's recommendation for $\frac{k_2}{k_1} = 0.55$:

$$\frac{M}{bd^2} = \rho f_y \left(1 - \frac{k_2}{k_1} \rho \frac{f_y}{f_{ck}^{prism}}\right) \quad ...(10)$$

For the section compression zone failure, the neutral axis position was obtained by using the compatibility equation based on the assumption of linear strain distribution:

$$f_s = E_s \, \varepsilon_{c\,ul} \, \frac{d - x}{x} \quad ...(11)$$

$$\frac{x}{d} = \sqrt{\frac{pm}{k_1} + \left(\frac{pm}{2k_1}\right)^2} - \frac{pm}{2k_1} \quad ...(12)$$

where $m = \frac{E_s \, \varepsilon_{c\,ul}}{f_{ck}^{prism}}$.

In the early 1930s, Steuermann proposed a theory for the section tensile zone failure of RC rectangular beams and slabs that considered the tensile concrete contribution [31]. This approach was applied in Russian specifications and the concrete tensile stress, f_{ct}, was accepted to be equal to $0.1f'_c$. As for the section compression zone failure, the parabolic stress-block with a neutral axis at a depth of $\frac{3d}{4}$ was converted into an equivalent triangular block (Figure 6).

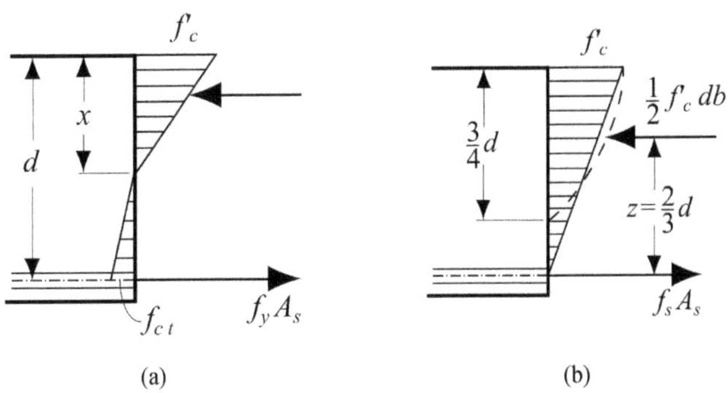

Fig. 6 Stress diagrams of a bent RC section, following [31]: (a) section tensile zone failure; (b) section compression zone failure.

The ultimate moment equation according to this approach is:

$$\frac{M}{bd^2} = \frac{1}{3} f_c' \qquad ...(13)$$

Another approach considering concrete contribution to tension due to shrinkage of concrete surrounding the reinforcement steel was proposed by Gebauer for the section tensile zone failure [32]. The main difference from the previously discussed theory is in the rectangular stress-block for compressed concrete area (Figure 7).

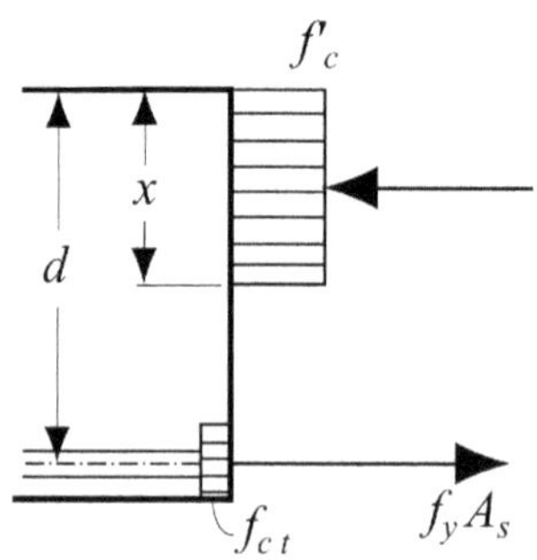

Fig. 7 Stress diagram of a bent RC section, following [32].

Later, Baumann came up with a parabolic type stress diagram for compressed concrete considering buckling of reinforced concrete columns under axial or eccentric loads [33] (Figure 8). The compressed concrete flexural strength was assumed to be equal to the prismatic one. It should be noted that testing eccentrically loaded prisms allowed to reveal the difference between the values of the ultimate concrete strain, $\varepsilon_{c\,ul}$, and that at axial compression when the concrete stress reaches its maximum value, ε_{c1}. For concrete class C20/25 ε_{c1} it was equal to 1.8‰ and $\varepsilon_{c\,ul}$ was 2.5...3.3‰. Thus, Baumann showed the importance of considering the strain hardening area in compressed concrete. Even though this phenomenon was discovered much earlier [14, 15], most researchers removed the measuring equipment before concrete failure, believing that failure occurs practically when the ultimate stress in compressed concrete is reached. Hereby, this stress-strain model for the compressed section zone is a basis for the bending analysis method used in modern codes, for example [2].

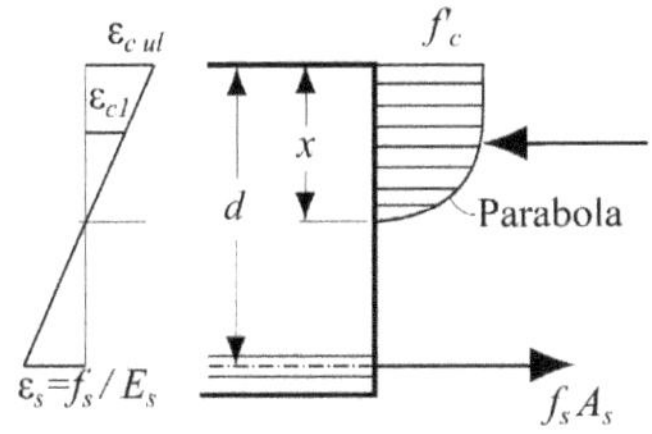

Fig. 8 Stress and strain diagrams of a bended RC section, following [33].

The same stress-strain block (Figure 8) was used by Bittner [34, 35] when the theory was based on Stüssi's approach [30]. The difference was in using ε_{c1} = 1.5‰ for all concrete classes. The rectangular stress-block with f_c' as the ultimate compressive stress for the section tensile zone failure was also proposed in this approach.

In 1935, Brandtzsæg presented a study including wide analysis of rectangular sections ultimate capacity based on testing 20 beams, 13 eccentrically loaded columns and numerous supplementary specimens [36]. This study included sections with and without compression reinforcement subjected to bending as well as to bending combined with axial load. The obtained results allowed to improve the Baumann's stress-block (Figure 8) still based on the Stüssi's approach for the section compressed zone failure and to obtain the following equation:

$$\varepsilon_{c\,ul} = \left(6.88 - 0.77\frac{f_c'}{1000}\right)10^{-3} \qquad ...(14)$$

Thus, the relationship between the ultimate concrete strain and compressive strength was obtained empirically: $f_c' = 0.85 f_c$. Additionally, plasticity ratio $\eta = \dfrac{\varepsilon_{c\,ul}}{\varepsilon_{c1}}$ was introduced and the relation between ε_{c1} and f_c' was also obtained. It allowed us to improve equations (8, 9, 11, and 12) by deriving the plasticity ratio, η instead of coefficients k_1 and k_2. Due to considering the compressed reinforcement and combined bending and axial load, equations (8 and 9) additionally changed.

For the beam's section tensile zone failure, Brandtzsæg used Suenson's equation (5) based on the rectangular stress-block. An equation for rectangular sections subjected to combined bending and axial load was also provided.

Finally, a modified standard procedure was developed, where a fairly uniform safety factor was achieved by introducing an allowable concrete compressive stress, which is a function of the load eccentricity. Emperger assumed that sufficiently accurate ultimate analysis of reinforced concrete beams may be carried out according to the proposed in Figure 9 models [37]:

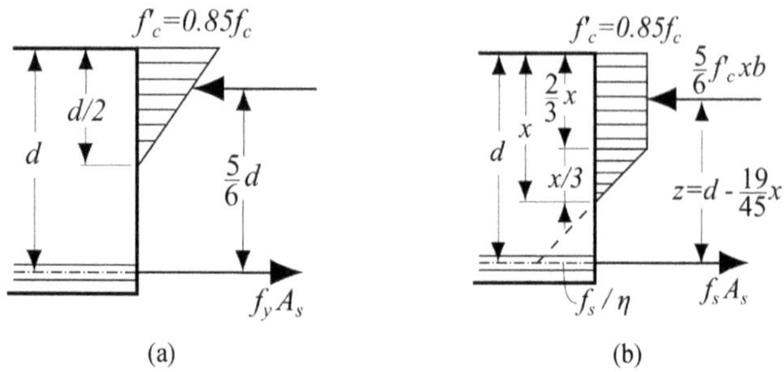

Fig. 9 Stress diagrams of a bent RC section, following [37]: (a) section tensile zone failure; (b) section compressed zone failure.

Saliger considered not only the compression and tension failures of rectangular section beams, but all five failure models [38]. Based on the Stüssi's approach, it was assumed that $k_2 = 0.5k_1$. It means that it was decided to use the rectangular stress-block with a depth of k_1x and $f'_c = f_c$. For the beams' section compression failure $\varepsilon_{c\,ul} = 3\ldots7‰$ was experimentally obtained. As a result, the following relationship was proposed:

$$\varepsilon_{c\,ul} = 1.75 f_c \cdot 10^{-6} \qquad ...(15)$$

where f_c is in psi units.

Whitney analyzed the ultimate states of reinforced concrete without considering ultimate strains, Bernoulli's hypothesis and the absence of slip. For the section tension failure the rectangular stress-block was provided (see Figure 2), where $f'_c = 0.85\,f_c$ was used for interpretation of the relationship proposed by Suenson earlier (5):

$$\frac{M}{bd^2} = \rho f_y \left(1 - \frac{1}{2}\rho \frac{f_y}{0.85 f_c}\right) \qquad ...(16)$$

For the compression section failure of beams Whitney proposed an equation similar to that of Steuermann (13), but based only on the test results, where $x/d = 0.537$:

$$\frac{M}{bd^2} = \frac{1}{3} f_c \qquad ...(17)$$

No compatibility equation was used in the study that completely distinguished this approach from Steuermann's one. For cross-section with compression reinforcement, the moment of full yield stress of the reinforcement was added to this equation. The equation for compression section zone failure under combined bending and axial loads was also proposed for rectangular sections with symmetrical reinforcement based mostly on an empirical data:

$$\frac{2A_s' f_y}{\frac{2e}{d'}+1}+\frac{bt\, f_c}{\frac{3te}{d^2}+1.178} \quad ...(18)$$

where A_s' is a cross-section area of the compressed reinforcement;
e is an eccentricity of the load application point relative to the section center of gravity.

The ultimate load for an axially loaded column can be obtained by substituting $e = 0$ into the equation (18). Substituting $e = \infty$ allows us to obtain the ultimate moment for a beam:

$$P = 0.85 f_c A_c + f_y A_s \quad ...(19)$$

A similar equation to that of Brandtzsæg was proposed by Whitney for the section tension failure in case of eccentrically loaded rectangular sections. In this equation the rectangular stress-block in the concrete was assumed.

In the Russian code, the stress-block in concrete was assumed as a cubic parabola for ultimate state design (Figure 10) [39, 40]. For the tension and compression failures of the rectangular beam cross-section, the equations were:

$$\frac{M}{bd^2} = \rho f_y\left(1-\frac{\rho f_y}{1.89 f_c'}\right) \quad ...(20)$$

$$\frac{M}{bd^2} = 0.4 f_c' \quad ...(21)$$

where f_c' is equal to 80 ... 100% of the ultimate strength of 20 cm (8 in) cubes.

The same assumptions were used for eccentrically loaded elements when the compression reinforcement is considered with the full yield-point stress.

For axially loaded columns, Russian codes proposed the following expression [39]:

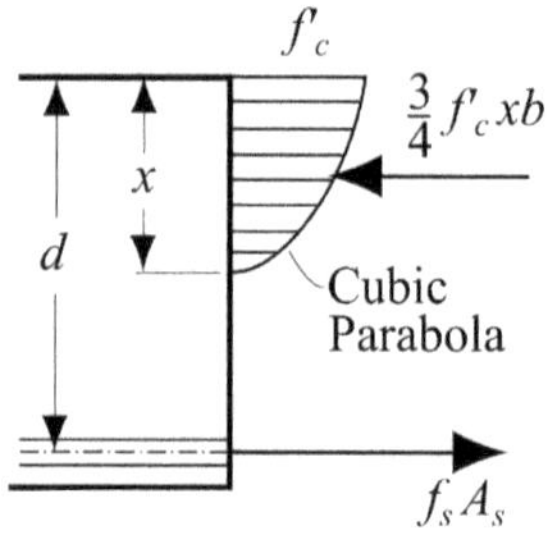

Fig. 10 Stress diagram of a bent RC section, following [39].

$$P = f_c A_c + f_y A_s \qquad ...(22)$$

where $f_c = \dfrac{4f_c'}{5}$.

In 1941, a theoretical conclusion based on experimental results for 110 rectangular section beams with and without compression reinforcement was presented by Cox [41]. The rectangular stress-block was chosen as the main theoretical scheme, when $f_c' = f_c$ (Figure 2). As a result, the same equation was obtained for the tension section failure as by Kazinczy [26]:

$$\frac{M}{bd^2} = \rho f_y \left(1 - \frac{1}{2}\rho \frac{f_y}{f_c}\right) \qquad ...(23)$$

An empirical value for the critical reinforcement ratio was determined for the compression section failure that was an intermediate value between Mensch's and Whitney's ones (1/2.6 and 1/3, respectively):

$$\rho_{cr} = 0.47 \frac{f_c}{f_y} \qquad ...(24)$$

$$\frac{M}{bd^2} = \frac{1}{2.76} f_c \qquad ...(25)$$

Later, Jensen represented his analysis of rectangular section beams without compression reinforcement [42, 43]. For the beam tension section failure, the method similar to Stüssi's was used based on a trapezoidal concrete stress-block with maximum stress $f_c' = f_c$ (Figure 11). An approach with a similar concrete stress-block was proposed earlier by Melan but not the cylindrical specimens' strength was applied [44].

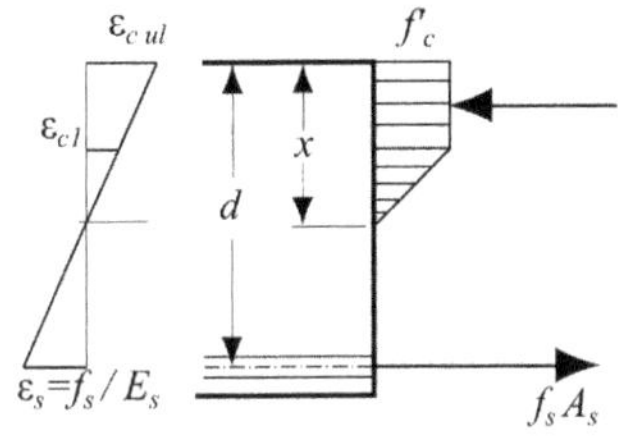

Fig. 11 Stress and strain diagrams of a bent RC section, following [42, 43].

Jensen described the strain when the stress reaches its maximum value according to Hooke's Law:

$$\varepsilon_{c1} = \frac{f_c}{E_c} \qquad ...(26)$$

where the concrete modulus of elasticity is:

$$E_c = \frac{30 \cdot 10^6}{5 + \dfrac{10000}{f_c}} \qquad ...(27)$$

The relationship between $\varepsilon_{c\,ul}$ and ε_{c1} was also provided:

$$\varepsilon_{c\,ul} = \frac{\varepsilon_{c1}}{1 - \eta} \qquad ...(28)$$

where η is a ratio considering plastic behavior:

$$\eta = \frac{1}{1 + \left(\dfrac{f_c}{4000}\right)^2} \qquad ...(29)$$

Finally, the equation for ultimate moment was obtained:

$$\frac{M}{bd^2} = \rho f_s \left(1 - \frac{1}{N}\rho\frac{f_s}{f_c}\right) \qquad ...(30)$$

where N is a function of η.

By assuming that $N = 2$ Jensen [42] optimized Kazinczy's equation (23) based on Stüssi's theory for the rectangular stress-block. The value of 2 was selected because in this case the ratio between Stüssi's parameters is $\dfrac{k_2}{k_1} = 0.5$. For the compression section zone failure, the compatibility equation and the trapezoidal stress-strain relation were used for the reinforcement steel.

A method for ultimate design of beams and calculation of the necessary reinforcement area in a rectangular section beam failing in tension under combined bending and axial load was developed by the Structural Bureau of the PCA [45], based on Jensen's approach [43]. A balanced section with additional compression reinforcement, reaching its full yield-point stress, was considered.

Chambaud also based the research on Stüssi's theory development for beams with rectangular cross-section and tensile reinforcement [46, 47]. A constant compressive ultimate strain for concrete cubes $\varepsilon_{c\,ul} = 3.6‰$ was obtained and also justified by experimental results. A concrete stress-block similar to that of Baumann, Brandtzsæg, and Bittner was first proposed (Figure 8) [46]. The initial modulus of elasticity as well as the maximum flexural stress was assumed to be equal to the values determined by testing 15 cm (5.5-in) cubes. Later, the stress-block type was modified from parabolic and linear parts to sinusoidal and parabolic ones (Figure 12) [47].

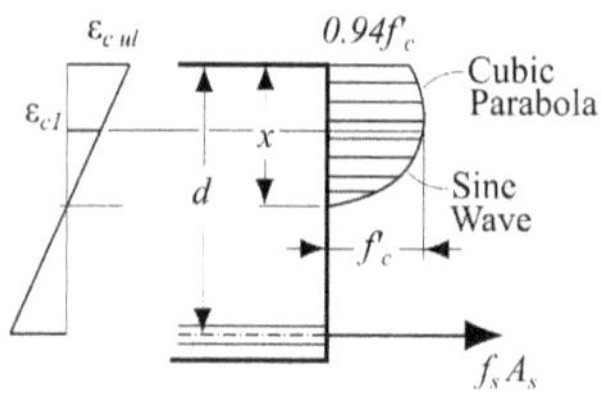

Fig. 12 Stress and strain diagrams of a bent RC section, following [47].

A general view of the equation for ultimate moments for the tension section failure type used by most scientists until the middle of the 20th century is:

$$\frac{M}{b\,d^2} = \rho f_y \left(1 - \frac{k_2}{k_1}\rho\frac{f_y}{f_c'}\right) \qquad ...(31)$$

The values used for the ultimate flexural compressive stress, f_c', were $(0.85\ldots1.18)f_c$, and for the $\frac{k_2}{k_1}$ ratio – 0.50 and 0.67 for rectangular and triangular stress distributions, respectively.

In the case of the compression section zone failure, most approaches were based on Stüssi's type theory:

$$\frac{M}{b\,d^2} = \alpha f_c \qquad ...(32)$$

where $\alpha = \frac{1}{2.6} \cdots \frac{1}{3}$.

Most of the researchers studied compression and tension failures of bent elements' sections. All of them considered the compression section zone failure by using two equations of equilibrium and one compatibility equation, which was used to obtain the strain relationships. In the case of the tension section failure, there was no need in the compatibility equation due to the initial assumption that tensile reinforcement reaches its yielding stress and concrete attains its ultimate stress.

According to this overview, in the middle of the 20th century the researchers concluded that it was necessary to base the structural design on the theory of inelasticity, as compressed concrete reaches inelastic state after about half of its bearing capacity. However, it was impossible to implement this approach without definition of stress-strain ($\sigma_c - \varepsilon_c$) relationship of compressed concrete from the beginning of loading until failure.

One of the first attempts to investigate $\sigma_c - \varepsilon_c$ relationship of compressed concrete was conducted at the beginning of the previous century [48]. The study included testing 10 × 20 cm concrete cylinders that allowed to obtain enough experimental data for stress-strain diagrams. Later, Hognestad investigated 190.5 cm-long compressed concrete column specimens with square sections of 25.4 × 25.4 cm and 223.5 cm-long ones with circular sections of 30.5 cm in diameter [4]. This research included testing 120 specimens with different eccentricity aimed at developing the inelasticity theory of concrete. Finally, an analytical parabolic model was proposed to describe the ascending branch of the diagram, and the descending one was represented by a linear dependence (Figure 13) [4].

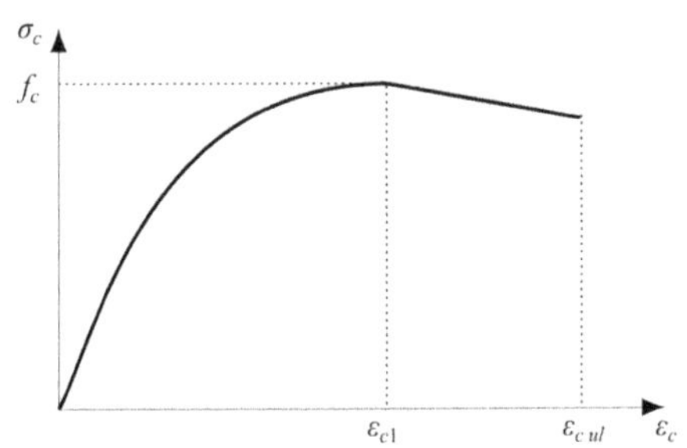

Fig. 13 Stress-strain dependency for compressed concrete, following [4].

The equation describing the initial part of this curve until ultimate concrete stress is presented in the form of a second-degree parabola:

$$\sigma_c = f_c \left[2\frac{\varepsilon_c}{\varepsilon_{c1}} - \left(\frac{\varepsilon_c}{\varepsilon_{c1}} \right)^2 \right] \qquad ...(33)$$

where σ_c is concrete stress;

ε_c is concrete strain.

The model was later modified by using new experimental data, obtained for ordinary concrete cylindrical specimens with a height-to-diameter ratio of 2. The specimens were subjected to compressive loading, controlled by a constant deformation rate [49, 50]:

$$\sigma_c = f_c \frac{\varepsilon_c}{\varepsilon_{c1}} \cdot \frac{p}{p - 1 + \left(\frac{\varepsilon_c}{\varepsilon_{c1}} \right)^p} \qquad ...(34)$$

where p is an experimental coefficient, expressed by an approximate function of the normal weight concrete compressive strength $p = 0.4 f_c \cdot 10^{-3} + 1.0$.

Stress-strain dependency for the case when only f_c is given [51]:

$$\varepsilon_{c1} = k \cdot 10^{-4} \sqrt[4]{f_c} \qquad ...(35)$$

where coefficient k depends on the aggregate type and the test method.

The linear descending branch was also proposed in a more recent work, aimed at obtaining an accurate model of normal and lightweight concrete behavior in compression [52].

The empirical relationship between σ_c and ε_c for compressed concrete, presented earlier [50], was further investigated by Carreira and Chu [53]:

$$\sigma_c = \frac{\beta \frac{\varepsilon_c}{\varepsilon_{c1}} f_c}{\beta - 1 + \left(\frac{\varepsilon_c}{\varepsilon_{c1}} \right)^\beta} \qquad ...(36)$$

where β is a material parameter that depends on the shape of the stress-strain diagram;

$$\beta = \frac{1}{1 - \frac{f_c}{\varepsilon_{c1} E_c}} \qquad ...(37)$$

The proposed analytical model differs from that of Popovics [50] by parameter β, but still involves unknown parameters that are obtained empirically from specimen compressive tests.

Model (34) was further investigated using the factor r increasing with stiffness reduction [54]:

$$\sigma_c = \frac{r \dfrac{\varepsilon_c}{\varepsilon_{c1}} f_c}{r - 1 + \left(\dfrac{\varepsilon_c}{\varepsilon_{c1}} \right)} \qquad \text{...(38)}$$

where

$$r = \frac{E_c}{E_c - E_{sec}}; \; E_c = 5000\sqrt{f_c}; \; E_{sec} = \frac{f_c}{\varepsilon_{c1}}$$

where E_{sec} is secant modulus of elasticity of concrete.

Another series of experiments were conducted on compressed cylindrical specimens (15 cm diameter and 30 cm height) and prisms (15 cm × 15 cm × 40 cm) that allowed to obtain a polynomial dependence for different types of concrete [55]:

$$\frac{\sigma_c}{f_c} = A\frac{\varepsilon_c}{\varepsilon_{c1}} + B\left(\frac{\varepsilon_c}{\varepsilon_{c1}}\right)^2 + C\left(\frac{\varepsilon_c}{\varepsilon_{c1}}\right)^3 \qquad \text{...(39)}$$

where A, B and C are coefficients, evaluated by plotting the experimental results in non-dimensional form.

According to the modern design codes [1, 2], the $\sigma_c - \varepsilon_c$ relationship for normal strength compressed concrete classes up to C60 is a diagram that includes a linear part in the elastic zone and a convex parabolic curve in the non-elastic one (Figure 14). According to this relationship, normal strength concrete (NSC) is described as elastic-plastic material with three basic performance stages:

- close to linear elastic behavior;
- non-linear elastic-plastic behavior;
- plastic behavior at failure.

Following those codes, the elastic stage, where the ratio between stresses and strains remains constant, is limited by 40% of concrete compressive strength, f_c. The elastic-plastic stage continues until compressed concrete deformations, ε_{c1}, at peak stress, f_c, reach the value of about 2.2‰ (depends on concrete class (*Eurocode 2*, Table 3.1 [1]). Finally, the plastic stage is limited by nominal ultimate strain, $\varepsilon_{c\,ul}$, which

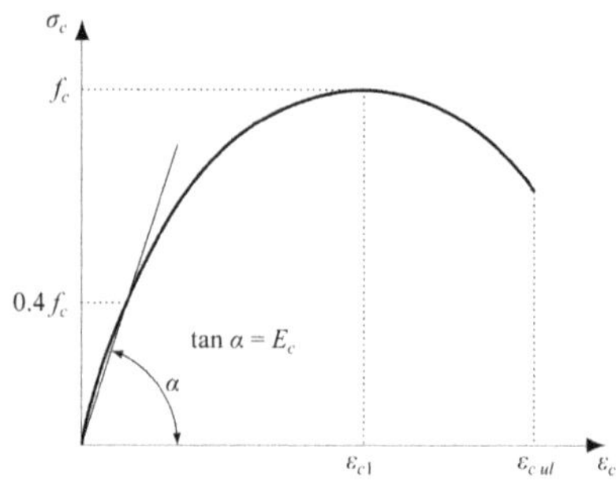

Fig. 14 Dependence σ_c vs. ε_c for compressed NSC, following [1].

is equal to 3.5‰ and determined as a constant value for NSC. From the mathematical point of view the relationship between σ_c and ε_c σ_c ε_c is determined as a convex square parabola [1]:

$$\sigma_c = f_c \frac{k\eta - \eta^2}{1 + (k-2)\eta} \qquad ...(40)$$

where

$$\eta = \frac{\varepsilon_c}{\varepsilon_{c1}}; \; k = 1.05 \frac{E_{\sec} |\varepsilon_{c1}|}{f_c}$$

The mathematical representation of concrete behavior in the non-elastic stage following [2]:

$$\sigma_c = 1000 f_c \ \varepsilon_c \ (1 - 250\ \varepsilon_c) \qquad ...(41)$$

In the case of sinusoidal interpretation of the same relationship, the following equation is obtained [56]:

$$\sigma_c = f_c \left(\frac{\pi \varepsilon_c}{0.004} \right) \qquad ...(42)$$

It should be mentioned that the difference in potentials in the case of using the square parabola or sinusoidal approximation is about 7% [56]. This, in turn, proves the validity of quadratic parabola expression for describing the stress-strain relationship of compressed NSC in a non-elastic behavior stage.

However, according to the described theoretical data there is a lack of several important parameters in the presented model:

- maximal elastic deformation value, $\varepsilon_{c\,el}$, which determines the limit of elastic stage;
- ultimate elastic deformation value, $\varepsilon_{c\,el\,ul}$, which is required for determining the full elastic concrete behavior potential;

- residual concrete stress that corresponds to the descending branch of the stress-strain diagram.

Additionally, it should be mentioned that the present model includes empirical parameters, which are determined as average values. For example, the limit stress of elastic stage is $0.4 f_{ck}$, concrete deformations at peak stress, ε_{c1} are proposed to be selected according to the table values given in [1].

There are a number of other analytical models that determine the stress-strain relationship. All these models are also based on empirical coefficients. As it can be seen from the proposed literature review [4], using relatively many empirical coefficients in the past is explained by the lack of theoretical justification of experimentally obtained relationships. However, the situation has not significantly changed until now. A model, which has experimental and theoretical significance, was recently proposed for simplifying the procedure, assuming such parameters as low axial strains, transverse deformations, and elastic behavior [57]. This method is an approximate one and can be easily implemented.

Analysis of available theoretical and experimental studies showed that the problem of using empirical coefficients to describe concrete behavior remained still relevant in recent years.

A theoretical stress-strain model for NSC design was proposed [3]. It enables the reduction of the number of empirical coefficients used in the design of concrete and RC structures. A brief description of the model is presented in the following subsection.

2.2 Basic Principles for the Theoretical Stress-Strain Model

The Theoretical stress-strain model for compressed concrete elements is based on the following three known principles: the Minimax Principle [58], Quasi-isotropic Concept of RC element at ultimate limit state (ULS) [59], and the Structural Phenomenon [60]. Each of these principles provides an additional equation (or a system of equations) for determining the behavior of RC elements under uniaxial compression, compression with eccentricity, and bending at ULS. This enables the designing of such elements without additional empirical coefficients. A summary of each of these three principles is presented below.

2.3 The Minimax Principle

The Minimax Principle idea is determined by obtaining the real load-bearing capacity of the structure, avoiding under- or over-estimation [58]. This becomes possible by selecting the compressed zone section depth (static parameter), which corresponds to the simultaneous implementation of the structure's maximum load-bearing capacity, when the external load value acting on the failure zone (kinematic parameter) reaches its minimum.

In the fourth decade of the 20th century, the ultimate equilibrium method was proposed by Gvozdev [61]. This approach was adopted for cement-based materials, and it is still used for the structural design at ULS. The limit state of a structure can be determined, on the one hand, as correspondence of internal forces to the plasticity conditions and the equilibrium equations in the section, and on the other hand, of reaching the external load limit value simultaneously. However, this method does not allow obtaining the exact value of load-bearing capacity due to separate consideration of the static or kinematic states.

The ultimate equilibrium method is based on three fundamental theorems: static, kinematic, and unity. According to the static theorem, the ultimate load should correspond to the statically permitted limit stress field, ψ, without considering element deformations (kinematic parameter). Thus, the lower bound of the structure ultimate loading capacity, P_u^-, is estimated. The kinematic theorem defines the ultimate load as the load causing the kinematically possible plastic deformation mechanism, r, when the stress condition compatibility is not considered (static parameter). In this case, the upper bound of the structure ultimate loading capacity, P_u^+, is obtained. In the second theorem, an idea of the key approach for combining the kinematically possible and the statically allowed states was proposed without any method for its solution [61]. However, the solution was necessary since each of the two theorems leads to obtaining results with certain deviations from real load-bearing capacity value [58]. Finally, the unity theorem enables us to obtain the ultimate loading capacity directly, but does not allow finding the maximum static or minimum kinematic loads.

As a result, the Minimax Principle became a logical development of the concrete elements' theory and proposed a solution for the above-described problem. The principle provides a solution for the unity theorem of the limit equilibrium method by combining the static and kinematic approaches in the same calculation, which means using both extreme features of the failure load. At the same time, only one method is used in calculations according to the limit equilibrium method – kinematic or static.

2.4 Quasi-isotropic Concept at Ultimate Limit State

The essence of the Quasi-isotropic idea is that a bent reinforced concrete element section at ULS is calculated assuming that it is made of a non-linear isotropic material [59]. This approach enables one to obtain an effective section with a minimum reinforcement ratio while the compressed zone depth reaches its maximum value, equal to half of the effective section depth [62].

If a section is made of a non-linear isotropic material, the neutral axis passes through the section center of gravity that is explained by ensuring internal forces equilibrium (Figure 15). Analytical results for a maximum moment, considering effective rectangular stress distribution, M_{max}, distance from the neutral axis to a layer with maximum stress, y_{extr}, and stress in the section top layer, $\sigma(h/2)$, for the case of square parabola and sinusoidal models, respectively [56], are:

$$M_{max} = 0.222bh^2 f_c;\ y_{extr} = 0.375h;\ \sigma(h/2) = 0.788f_c \qquad ...(43)$$

$$M_{max} = 0.218bh^2 f_c;\ y_{extr} = 0.377h;\ \sigma(h/2) = 0.754f_c \qquad ...(44)$$

The obtained results prove that any type of concave curve is suitable for structural analysis.

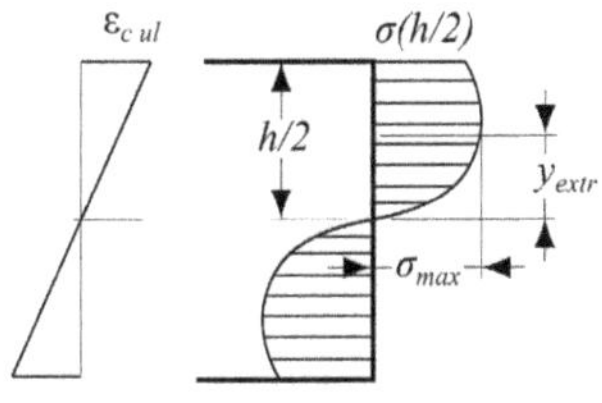

Fig. 15 Stress and strain diagrams for non-linear isotropic material bent section, following [56].

For a section made of non-linear non-isotropic material, such as pure concrete, the neutral axis still passes through the center of gravity. It is proved by using the Structural Phenomenon (that will be explained in the Section 2.5) that the concrete modulus of elasticity decreases twice when the concrete reaches its limit tensile strength [56]. Moreover, it is justified based on a wide experimental data that the neutral axis depth of effectively designed RC rectangular section cannot exceed a half of the effective section depth, otherwise, an over-reinforced section will be obtained that leads to the brittle collapse case:

$$x \leq 0.5d \qquad (45)$$

In a case when the compressed section depth $x = 0.5d$, the reinforced concrete can be determined as a quasi-isotropic material since it has the same behavior characteristics at ULS as an isotropic material. According to the standard concrete theory interpretation, the non-rectangular section maximal compressed depth can be calculated by using the following expression:

$$S_{c,\max} = 0.75\, S_0 \quad \text{...(46)}$$

where $S_{c,\max}$ is static moment of the compressed section zone, relative to the tensile reinforcement;

S_0 is the static moment of the section effective zone.

At this design stage there are only two equilibrium equations (moments and forces) with three unknowns (cross-section area of compressed and tensile reinforcement and neutral axis depth). For theoretical justification of conditions (44) and (45) an additional equation was proposed, based on the idea of an effective reinforcement use [56]:

$$A_s + A'_s \rightarrow \min \quad \text{...(47)}$$

Then:

$$\frac{d(A_s + A'_s)}{dx} = 0 \left(\frac{d^2(A_s + A'_s)}{dx^2} > 0 \right) \quad \text{...(48)}$$

The obtained solution for the problem of a standard bent rectangular section (Figure 16) provides the following justifications:

- the cross section reaches its quasi-isotropic state when the total area of reinforcement section is minimal, since the minimum of the function $[(A_s + A'_s)(\omega)]$ corresponds to the maximal value of ω, when $\omega = x/d$;
- parameter ω does not depend on external moment value that corresponds to one of the main characteristics of isotropic material.

$$\frac{d(A_s + A'_s)}{d\omega} = -\frac{2d}{d - d'_s} + \frac{2d\omega}{d - d'_s} + 1 = 0 \quad \text{...(49)}$$

$$\omega = \omega_{extr} = 0.5\left(1 + \frac{d'_s}{d}\right) \quad \text{...(50)}$$

Theoretically, ω varies in the range of $\frac{2d'_s}{d} \leq \omega \leq \omega_{extr}$, when the lowest value ensures proper concrete cover layer for compression reinforcement. According to *Eurocode 2* [1], ω equals to 0.43 or 0.5, depending on the selected stress-strain diagram type; following [2] it is limited by the value of 0.4. The theoretical values for rectangular sections with both tensile and compression reinforcement can be $\omega = 0.5 \ldots 0.625$. In the case of using only tensile reinforcement, $\omega_{extr} = 0.5$.

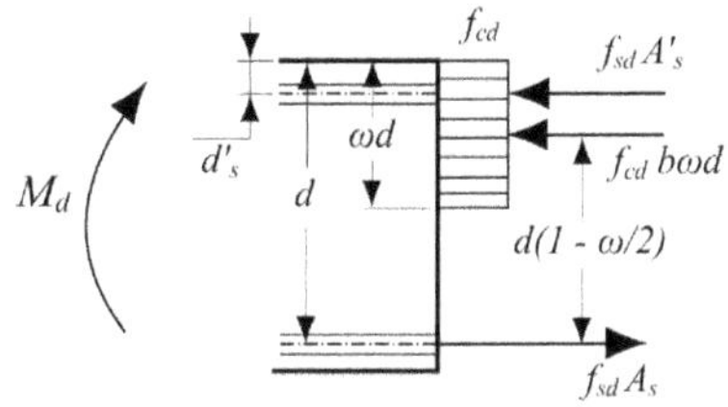

Fig. 16 Static scheme of a bent rectangular cross-section, following [56].

By investigation of the function $[(A_s + A'_s)\ (\omega)]$, it can be discovered that the cross-section area of the tensile reinforcement increases with a corresponding decrease in that of the compression one, as well as of the function $[(A_s + A'_s)\ (\omega)]$ values until reaching ω_{extr}. Further, behavior is described by a joint increase in the cross-section area of tensile reinforcement and values of the function; however, this part cannot be taken into consideration since ω cannot exceed the value of ω_{extr}. Thus, material can be called quasi-isotropic, when the following two conditions are satisfied [56]:

- reinforcement reaches its yielding in the tensile zone. Consequently, compressed concrete, as well as compression reinforcement, on a material level reach the plasticity stage and the section reaches its load-bearing capacity;
- neutral axis depth for rectangular section is equal to half of the effective cross-section depth $x = 0.5d$ (for non-rectangular sections the neutral axis passes through the section center of gravity).

As a result, an additional equilibrium equation was obtained. This excludes the need for empirical coefficients, similar to the case when the Mini max Principle was applied.

2.5 Structural Phenomenon

The Structural Phenomenon was proposed as a result of extensive experimental and theoretical data analysis of RC structures' behavior at three levels: material level, static level, and dynamic loadings level [60]. The phenomenon's essence is that parameters of the structure or its element at ULS increase or decrease twice, compared to the corresponding values in the initial and elastic states. This phenomenon was proved for symmetrical sections of structural elements subjected to symmetrical loading. Using the Structural Phenomenon, as well as the above-described principles (Mini max, Quasi-isotropic), provides additional equations that also eliminate the empirical coefficient.

Some examples for the above-mentioned effects on the RC structure performance (following [60]):

I. Material level

a. Comparison of maximal concrete tensile strains for C20 and C70 concrete classes, according to [62], shows the twofold difference between the values:

$$\frac{e_{ctm}\ C70}{e_{ctm}\ C20} \approx \frac{0.12\%}{0.06\%} = 2 \qquad ...(51)$$

b. The mid-span displacement, Poisson deformations, and horizontal shear deformations between the concrete layers under the ultimate load, Δ_{ul}, $\varepsilon_{trans\ ul}$ and $\varepsilon_{shear\ ul}$, and after unloading, Δ_{res}, $\varepsilon_{trans\ res}$ and $\varepsilon_{shear\ res}$, were experimentally investigated for a full-scale 3 m two-layer beams tested under four-point loading [63]:

$$\frac{\Delta_{ul}}{\Delta_{res}} = \frac{35\ \text{mm}}{18\ \text{mm}} = 1.94 \approx 2 \qquad ...(52)$$

$$\frac{\varepsilon_{trans\ ul}}{\varepsilon_{trans\ res}} = \frac{0.65‰}{0.32‰} = 2.03 \approx 2 \qquad ...(53)$$

$$\frac{\varepsilon_{shear\ ul}}{\varepsilon_{shear\ res}} = \frac{2.70‰}{1.25‰} = 2.16 \approx 2 \qquad ...(54)$$

All relations showed the results close to 2. All inaccuracies are considered acceptable for the experimental data for such non-isotropic material as concrete.

c. Experimental results for minimal and maximal Poisson coefficient values were obtained by loading cylindrical concrete specimens without steel fibers, ν_0 and v_0, and with the optimal fiber content of 30 kg/m^3, ν_{30} and v_{30} [60].

$$\frac{\nu_{30}}{\nu_0} = \frac{v_{30}}{v_0} = 2 \qquad ...(55)$$

$$\frac{v_{30}}{\nu_{30}} = \frac{v_0}{\nu_0} = 2 \qquad ...(56)$$

where $\nu_0 = 0.11$, $\nu_{30} = 0.22$, $v_0 = 0.22$, $v_{30} = 0.44$.

A high convergence of the presented experimental data with the Structural Phenomenon was received.

II. RC element behavior under static loading.

a. *Uniformly distributed load.* Maximal elastic and plastic moment values, M_{el} and M_{pl}, for fixed-ended beam (Figure 18) under uniformly distributed load, q:

$$M_{el}^{+} = \frac{q_{el}\, l^2}{24};\ \left|M_{el}^{-}\right| = \frac{q_{el}\, l^2}{12} \qquad ...(57)$$

$$M_{pl}^{+} = \left|M_{pl}^{-}\right| = \frac{q_{ul}\, l^2}{16} \qquad ...(58)$$

Then the Structural Phenomenon evidence:

$$M_{pl}^{+} + \left|M_{pl}^{-}\right| = M_{el}^{+} + \left|M_{el}^{-}\right| \frac{ql^2}{8} \qquad ...(59)$$

b. *Bending moment.* This example was described in subsection 2.4, where a non-isotropic RC section reaches a quasi-isotropic state at ULS when the optimal reinforcement condition is achieved. Additionally, the concrete compression zone depth is equal to half of the effective section height. Consequently, the Structural Phenomenon corresponds to the optimal design concept of cement-based composite bending elements and structures.

Moreover, in the case of using compressed reinforcement in a section, its load-bearing capacity is determined as a sum of maximal values of moments transferred to compressed reinforcement, $\Delta M_{\max}$, and concrete, $M_{cd\max}$:

$$M_{d\max} = \Delta M_{\max} + M_{cd\max} \qquad ...(60)$$

It is known, that at ULS $\Delta M_{\max} = M_{cd\max}$, then:

$$M_{d\max} = 2M_{cd\max} \text{ or } M_{d\max} = 2M_{sd\max} \qquad ...(61)$$

Thus, the quasi-isotropic section load-bearing capacity can be increased by a factor of two at the maximum if compression reinforcement is used in the section.

c. *Shear force.* The ultimate shear resistance of an RC element is determined as a sum of concrete and reinforcement maximal forces, V_{Rdc} and V_{Rds} [1, 2]:

$$V_{Rd\max} = V_{Rdc} + V_{Rds} \qquad ...(62)$$

In the ultimate case:

$$V_{Rd\max} = 2V_{Rdc} \quad ...(63)$$

since:

$$V_{Rds} = V_{Rdc} \quad ...(64)$$

III. RC structure and element behavior under dynamic loading.

a. *Full-scale three-story RC frame.* A full-scale three-story beamless precast framed RC building part was tested under impulse step loading [64]. The dominant natural vibration period of the frame before testing was $T_{1\,\min} = 0.6$ sec, and after transferring it to the non-linear system it was $T_{1\,\min} = 1.2$ sec. That means a double increase from the elastic to the plastic state corresponding to the Structural Phenomenon principle.
b. *Base isolation system for a 9-storey building.* A dominant natural vibration period equal to 0.72 sec was experimentally obtained for a full-scale 9-storey RC building designed for an earthquake zone with a peak ground acceleration (PGA) equal to 0.15 g and a corresponding base shear force (BSF) of 3400 kN [65]. After applying a base isolation system, designed for PGA equal to 0.3 g, the dominant vibration period reached the value of 1.4 sec when BSF practically remained the same. Thus, it was proven that increasing the dominant vibration period of the building about twice allows its adaptation to a zone with a twice higher seismicity without any significant change in the building response.

2.6 Combine Application of Structural Phenomenon and Minimax Principle

Two dependencies for a 1 : 10 scaled model of 24 × 24 m RC shell performance are presented in Figures 17 and 18. The first one represents the relation between the dominant vibration mode frequency, f_l, and load ratio, q/q_{ul}. There is experimental evidence that the frequency decreased twice when the model reached the ULS. Figure 18 shows the dependence of residual deflections, Δ_{res}, on various damping ratios, ξ, of the same shell model under static and additional vertical seismic loading. Records of real earthquakes with high-, medium-, and low-frequency ground motions were taken as examples of dynamic loading. The next level of displacement values were obtained:

- 2.2 cm at static loading;
- from 2.2 to 4.4 cm at a high-frequency earthquake;

- from 2.0 to 3.6 cm at a medium-frequency earthquake;
- from 2.0 to 2.5 cm at a low-frequency earthquake.

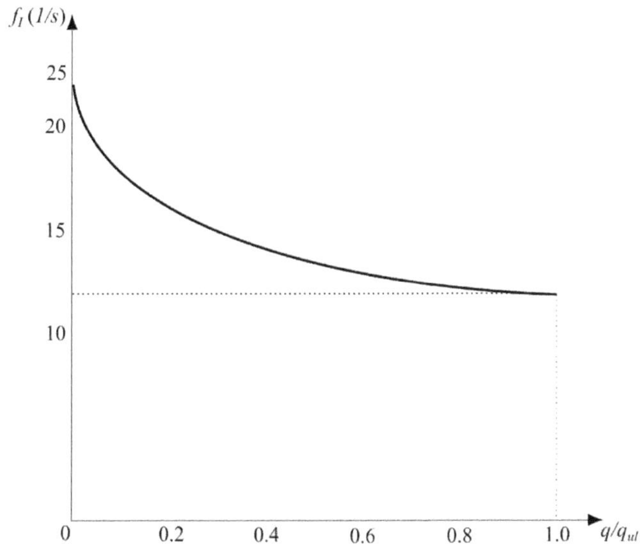

Fig. 17 Relation between the dominant vibration mode frequency and load ratio, following [60].

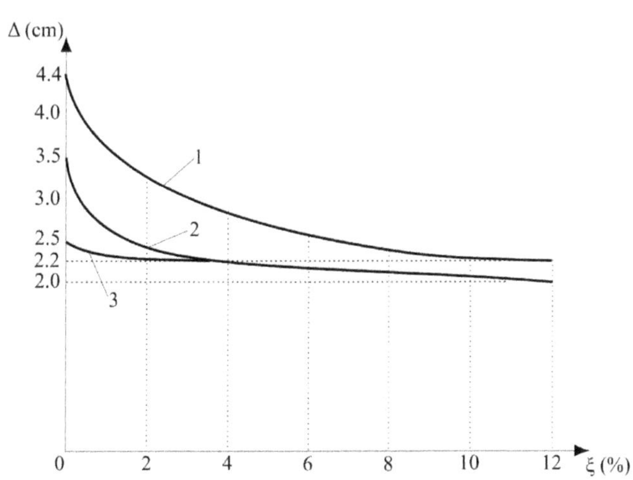

Fig. 18 Relation between shell residual deflections and damping ratios, following [60]: 1. High-frequency earthquake; 2. Medium-frequency earthquake; 3. Low-frequency earthquake.

As a result, for high- and medium-frequency dynamic loads the residual deflections doubled proving the Structural Phenomenon. The Minimax Principle interpretation can be demonstrated based on the presented results (Figure 19). The Principle combines the curves from Figures 17 and 18 as the ones in planes n-0-q/q_{ul} and m-0-q/q_{ul}, respectively, and presents their structural interrelation in the n-0-m plane [60]. Consequently, in case of some structural parameter double rise at ULS (axis m), justified by empirical data, a corresponding double reduction can be prognosed for another parameter (axis n), of the observed structure.

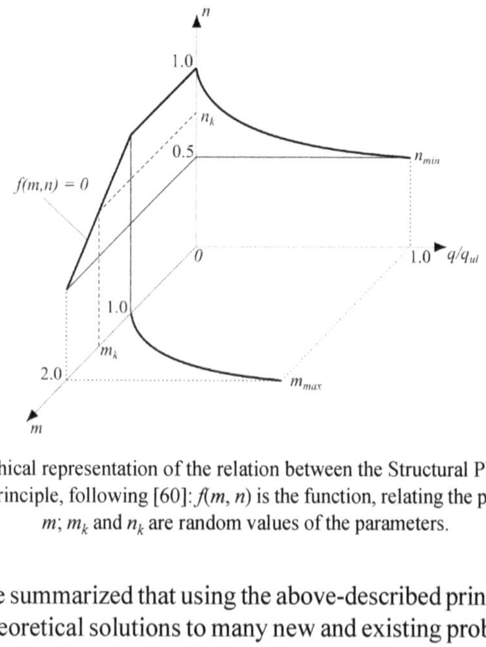

Fig. 19 Graphical representation of the relation between the Structural Phenomenon and the Minimax Principle, following [60]: $f(m, n)$ is the function, relating the parameters n and m; m_k and n_k are random values of the parameters.

It can be summarized that using the above-described principles allows obtaining theoretical solutions to many new and existing problems, which could be previously solved only by using empirical coefficients. Thus, further development of the limit equilibrium method and the general theory of RC elements became possible.

2.7 Theoretical Stress-Strain Model for Compressed Concrete

The three above-described principles (Mini max, Quasi-isotropic, and Structural Phenomenon) enabled the formulation of the main provisions for developing a theoretical model of stress-strain relationship in compressed concrete (Figure 20) [3]. The novelty of this model was the theoretical approach for finding the stress-strain diagram parameters without applying empirical coefficients. The main parameters of this diagram for concrete classes up to C60 are:

- maximum value of elastic deformations $\varepsilon_{c\,el}$ = 0.5‰ (see point 1 in Figure 20);
- limit concrete elastic potential determined by the deformation $\varepsilon_{c\,el\,ul}$ = 1 (this potential is generally required for structural analysis under dynamic loads and for concrete creep analysis) (see point 2 in Figure 20) [66];
- ultimate elastic stress reaching half of the concrete strength, i.e., $\sigma_{c\,el} = 0.5f_c$ (see point 1 in Figure 20);
- boundary of the diagram descending branch is determined by the stress equal to $\sigma_{c\,ul} = 0.5f_c$ (it helps to find the real value of the compressed concrete cross-section potential) (see point 4 in Figure 20).

Other parameters determining compressed concrete behavior in longitudinal direction are defined as in modern codes [1, 2]:

ε_{c1} is compressive strain in concrete at peak stress, equal to 2‰ (see point 3 in Figure 20);

ε_{cu} is the ultimate compressive strain in concrete, equal to 3.5‰.

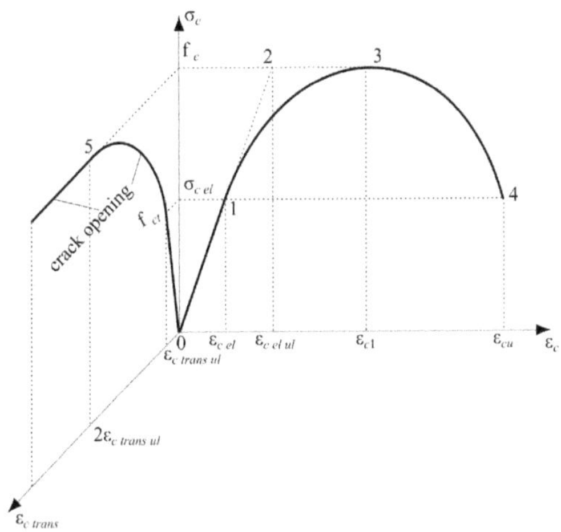

Fig. 20 Compressed concrete stress-strain diagram: 1–the limit of concrete section elastic stage; 2–limit elastic potential of a section; 3–concrete strength; 4–the final point of the descending branch; 5–the border between non-linear transverse deformation and compressed concrete ULS in a longitudinal direction.

Compressed concrete specimen behavior in transverse direction is also presented in Figure 20 (see $\sigma_c - \varepsilon_{c\ trans}$ axis). According to the diagram, the following parameters are used:

f_{ct} is the concrete tensile strength in the transverse direction under longitudinal compression, corresponding to the appearance of longitudinal cracks in the specimen;

$\varepsilon_{c\ trans\ ul}$ is the ultimate elastic transverse strain, determined by Poisson's coefficient and equal to 0.1‰ [15], which is approximately 20% of the corresponding longitudinal deformation value.

A detailed description of the main stages in transverse deformation development is given in Chapter 3.3.

Additionally, concrete elastic and plastic potentials, U_{el} and U_{pl}, can be calculated based on the presented stress-strain relationship. The elastic one can be found as [3].

$$U_{el} = 0.5 \cdot 0.5f_c \cdot 0.5 + 2 \cdot 0.5 \cdot 0.5f_c \cdot (2 - 0.5) = 0.875f_c \qquad ...(65)$$

and the plastic one as:

$$U_{pl} = \frac{2}{3} \cdot 0.5f_c \cdot (3.5 - 0.5) + 2 \cdot 0.5f_c \cdot (2 - 0.5) = 1.75f_c \qquad ...(66)$$

Another justification of the Structural Phenomenon is provided in this case as follows:

$$\frac{U_{pl}}{U_{el}} = \frac{1.75f_c}{0.875f_c} = 2 \qquad ...(67)$$

3

Transverse Deformations as One of the Basic Features of Compressed Concrete Element Performance

3.1 Influence of Relatively High Loading Rates on Transverse Deformations

The transverse deformation analysis of structures is usually limited to obtaining the Poisson's ratio, which is only applicable to the range of elastic concrete behavior. At the same time, for compression, this range can reach half of the concrete strength. Consequently, the elastic compressed concrete stage is an essential part of concrete behavior investigations [3].

One of the important factors affecting compressed concrete behavior is the specimen's loading rate. Following the requirements of modern codes, it is recommended to load a concrete specimen at a rate within (0.2 ... 1.0) MPa/s [67]; according to [68] the recommended rate is (0.25 ± 0.05) MPa/s. However, at such loading rates, longitudinal deformations, even for NSC, mainly exhibit elastic behavior. Some examples proving this effect are presented below.

Twenty-two cylindrical NSC specimens were tested under uniaxial loading [69]. The specimen diameters were between 10 and 80 cm, and their heights were from 20 to 320 cm (which corresponds to a real column length). The specimens were loaded at a rate of about 0.1 MPa/s. The obtained $\sigma_c - \varepsilon_c$ relationships, (σ_c was the compressed concrete stresses and ε_c was the longitudinal deformations), were characterized by an almost linear performance up to the specimens' failure. Similar results were obtained for uniaxial loaded 15 × 30 cm cylindrical specimens made of steel fiber reinforced concrete with a strength of 27.6 MPa [70]. In this case, the loading rate was 1.3 mm/min (which corresponds to about

0.1 MPa/s). Comparable results were obtained for 26 cylindrical NSC specimens loaded at a rate of (0.25 ± 0.05) MPa/s [71].

The dominant elastic behavior almost until failure is mostly typical for high-strength concrete (HSC) that represents classes C70 … C90 [1]. It is known that HSC exhibits mainly elastic behavior with weakly expressed elastic-plastic deformations [72]. For more accurate identification of HSC non-elastic deformations, the use of transverse deformation graphs of the compressed specimens was proposed [73]. It was shown that in certain cases, particularly in steel-fibered high-strength concrete (SFHSC), transverse deformations efficiently indicate elastic-plastic concrete behavior, which is less pronounced when longitudinal deformations are used. Experimental results obtained by testing 14 cylindrical specimens with 15 cm diameter and 30 cm height subjected to static uniaxial loading confirmed the importance of using transverse deformations for analysis of compressed concrete behavior [73].

However, this approach is not suitable for regular HSC. First of all, longitudinal inelastic deformations in compressed HSC specimens are practically not manifested as well. Secondly, the value of transverse deformations is approximately five times less than that of the longitudinal ones (Poisson's ratio is approximately equal to 0.2). It means that they also do not provide a good solution for the HSC problem since the initiation of longitudinal cracks caused by transverse deformations practically coincides with specimen failure.

For this reason, another solution was proposed by the authors [74]. As mentioned above, NSC behaves elastically almost until failure at rather high loading rates, as HSC. However, maximal longitudinal deformations in NSC are almost twice as high as in HSC. At the same time, concrete tensile strength practically does not grow when the concrete class increases to over C70. Therefore, transverse deformations development in NSC is much more significant than in HSC, which leads, in turn, to the longitudinal crack development that determines inelastic concrete behavior in the specimen transverse direction. Thus, NSC subjected to relatively high loading rates can be considered as a model of HSC behavior at normal loading rates based on transverse deformation analysis.

3.2 State-of the Art Approach of Transverse Deformation Application for Concrete Limit States Analysis

Although many experimental investigations were carried out to study concrete specimen behavior in longitudinal direction, few of them are focused on transverse deformations over the elastic limit.

Sixty cylindrical specimens with 10 cm diameter and 20 cm height, made of self-compacting NSC with fine and coarse recycled concrete aggregate, were subjected to a monotonic loading, as well as to loading-unloading tests with increasing the load magnitude [75]. The obtained experimental data demonstrates a significantly more pronounced inelastic behavior of specimens in transverse direction, which is evident from the stress-transverse deformation diagrams.

Using circumferential strains for deep analysis of a rebar pull-out test from concrete specimens enabled enhanced understanding of bonding resistance mechanism [76]. For this purpose, 12 cylindrical specimens with diameters of 100, 150, and 200 mm and height of 200 mm were tested. The obtained experimental results allowed developing a new theoretical three-dimensional model of the specimen behavior during a pull-out test.

An experimental study on transverse deformations in prestressed reinforced concrete (RC) two-layer beams (TLB) with a span of 8 m was conducted [77]. One of the aims of this study was improving the concrete theory related to the Poisson's ratio in elastic and inelastic stages. The research was focused on concrete behavior in the compressed zone of bending RC elements. As a result, a new limit state for such elements was obtained. It allows a more accurate prediction of cracks' development. This research was preceded by a long-term experimental program which included investigation of 0.7 m long single-supported, 3 m long single-supported, continuous 4 m long two-span, and pre-stressed 3 m long TLBs. It should be noted that in all these studies transverse deformations were considered as valuable indicators of effective concrete performance.

It is logical to apply experimental longitudinal and transverse deformations in a step-by-step analysis that enables one to determine a limit state of compressed concrete cylindrical specimen by the crack depth which is a function of crack width. Thus, a new approach for limit-state analysis of compressed concrete elements using transverse deformations is proposed.

3.3 Transverse Deformation Development in Compressed Concrete

Available literature on initiation and development of cracks in compressed concrete elements is rather limited. As one of the first attempts to fill this gap, a theoretical model of the longitudinal cracks' development in compressed concrete cylindrical specimens, including consequent scheme for specimen failure was proposed [73]. This model allows predicting cracks' development during the entire loading process and includes the following three main stages (Figure 1):

- linear specimen behavior;
- non-linear specimen behavior;
- ultimate limit state.

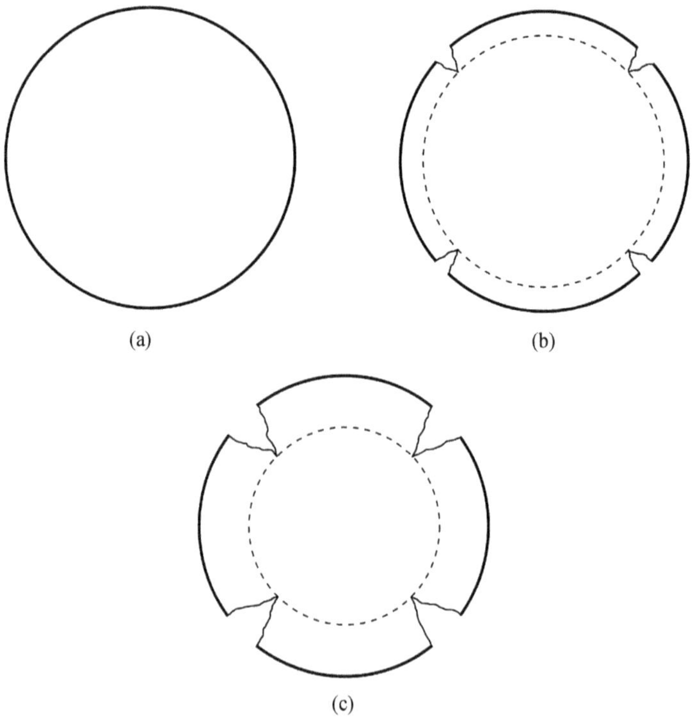

Fig. 1 Compressed cylindrical concrete specimen performance considering transverse deformations: (a) before transverse cracking ($\varepsilon_{trans} \leq \varepsilon_{ct\,ul}$); (b) during cracking ($\varepsilon_{ct\,ul} < \varepsilon_{trans} \leq 2\varepsilon_{ct\,ul}$); (c) at ULS ($\varepsilon_{trans} > 2\varepsilon_{ct\,ul}$), following [73].

The first stage continues until the initiation of cracks ($\varepsilon_{trans} \leq \varepsilon_{ct\,ul}$), where ε_{trans} is the transverse deformation value, and $\varepsilon_{ct\,ul}$ is the ultimate elastic transverse deformation. At this stage, the cylindrical specimen's cross-section is non-damaged and has an original circular shape. The second stage begins from the initiation of cracks, corresponding to $\varepsilon_{ct\,ul}$, and continues until a twofold ultimate elastic transverse deformations value ($\varepsilon_{ct\,ul} < \varepsilon_{trans} \leq 2\varepsilon_{ct\,ul}$) is reached, based on the Structural Phenomenon concept [60]. In this case, it is assumed that the specimen cross-section has an inner circular part (unbroken core) and several outer wedge-shaped segments separated by longitudinal cracks, that is the result of their depth increase under the external loading. The core-section area before failure is equal to a part of the initial total cross-sectional one and can be also calculated using the Structural Phenomenon that determines the compressed concrete element limit state in the longitudinal direction. The last stage corresponds to transverse deformations that exceed $2\varepsilon_{ct\ ul}$. It was theoretically predicted that after reaching the ULS the area of the specimen inner core is approximately half of the initial value. Similar concrete and mortar cylindrical specimen failure patterns were also obtained under cyclic uniaxial compression [78].

As it was mentioned above, Figure 20 (Chapter 2) presents a compressed concrete performance in both longitudinal and transverse directions. It shows the relationship between the following three dependencies:

- longitudinal strains, $\varepsilon_{c\ long}$;
- transverse strains, $\varepsilon_{c\ trans}$; and
- compression stresses, σ_c.

Actually, it is a 3D diagram represented by two projections for better clarity. According to this Figure 20 (Chapter 2), the non-linear concrete behavior in transverse direction is presented by the ascending branch of the diagram that starts from f_{ct} and continues up to point 5, corresponding to $2\varepsilon_{c\ trans\ ul}$ deformation value (Figure 20 [Chapter 2], $\sigma_c - \varepsilon_{c\ trans}$. This stage determines the longitudinal crack opening process. After reaching the concrete compressive strength, f_c (see points 3 and 5 in $\sigma_c - \varepsilon_c$ and $\sigma_c - \varepsilon_{c\ trans}$ diagrams, respectively), further crack development corresponds to specimen failure process (descending branch in longitudinal stresses, $\varepsilon_{c\ el} = 2.0 - 3.5‰$).

4

Test Series Coverage

4.1 Research Stages and Testing Methodology

A total of six separate test series were conducted for obtaining a valuable experimental basis. The number of specimens in a series was limited by the mixer volume, which was used for preparing the concrete mixture (0.045 m^3). Therefore, each series included three cuboids with a rib size of 15 cm and six cylindrical specimens (15 × 30 cm), excluding the first research stage, where a ready-mix concrete was used. However, experimental data from fewer specimens was used in the analysis process due to manufacturing defects in some specimens. This is explained by the complexity of concrete as a material due to its heterogeneity and non-isotropy.

4.1.1 The Preliminary Research Phase Description

The first research phase was preceded by two series of experiments: the preliminary research phase, consisting of a total of six cubic and eight cylindrical specimens that was aimed at working out the selection of the concrete mix composition, as well as the initial adjustment, calibration, and debugging of the measuring equipment (Figure 1).

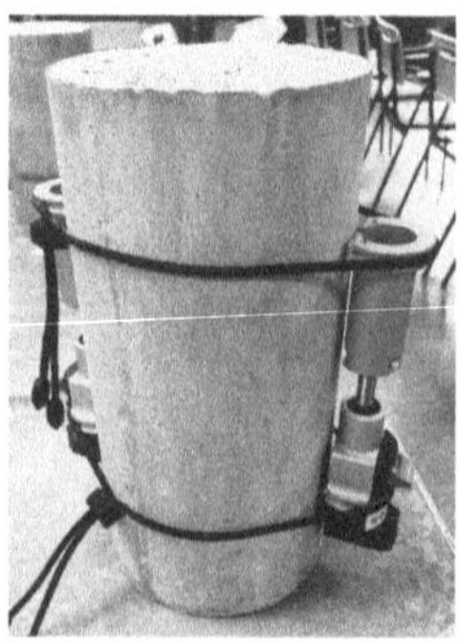

(a) (b)

Fig. 1 Specimens from the first two preliminary test series: (a) cubic; (b) cylindrical.

The concrete mixture composition was adjusted based on the obtained concrete strength as well as on the fresh mixture properties. The distribution of aggregates in a specimen body was also analyzed after its fracture (Figure 2). A non-uniform distribution of coarse aggregate was found in cylindrical specimens. It was revealed that improper external vibration caused the aggregate grains to sink down and left the upper part of a specimen free of them. In the second series, this problem was solved by using internal vibration for a limited time (see paragraph 4.1.3) (Figure 2c).

(a)

(b) (c)

Fig. 2 Internal specimen structure after failure: (a) cubic specimens; (b) cylindrical specimen from the first preliminary test series; (c) cylindrical specimen from the second preliminary test series after vibration approach correction.

Additionally, the second preliminary test series was aimed at developing the technique of leveling the specimen upper surface using fine grain sand, as well as developing the skill of gluing strain-gauge type sensors to the concrete surface (Figure 3). Moreover, the initial adjustment and calibration of the measuring equipment was carried out and the method of longitudinal sensor attachment to a specimen was mastered (a detailed description of the measuring equipment is presented in paragraph 4.6).

Fig. 3 Cylindrical concrete specimen with sifted fine grain sand on the upper surface and strain-gauge placed in the transverse direction.

4.1.2 The Main Research Phases

The research included the following two main phases. In the first one six cylindrical specimens, subjected to axial loading, were tested in order to obtain the experimental stress-strain relationship for NSC, including the descending branch. The measuring equipment suitability was preliminarily verified by applying step loading and unloading to three cylindrical specimens. For this reason, 10 load steps were selected (each load step was 40 kN). After that, two test series were conducted when the behavior of three specimens from each series was considered. The specimens were marked as Sp1.1, Sp1.2, Sp1.3, Sp2.1, Sp2.2, and Sp2.3, respectively. The concrete strength corresponded to C20/25 and C35/45 concrete classes, respectively [1, 79]. The lower-strength concrete was chosen to obtain a more pronounced specimen behavior in the non-elastic stage, including the descending branch.

In the framework of the second phase, the analysis of transverse deformations and their relation to longitudinal ones was of prime focus. The research included two test series where an analysis of elastic-plastic behavior of 15 × 30 cm cylindrical specimens under a constant loading rate of 0.2 MPa/s was performed. The first and second test series included the behavior analysis of four and three specimens, marked as Sp3.1, Sp3.2, Sp3.3, Sp3.4, Sp4.1, Sp4.2, and Sp4.3, respectively. For the new purposes of this research phase a significant change in the testing system was required. The measuring system was also completely changed that will be precisely described in paragraph 4.6.

4.2 Specimen Shape and Dimensions

Concrete strength was determined by testing 15 cm cubic specimens. Cylindrical specimens were chosen as the main type for conducting experiments in this research, because there is no local stress concentration problem as, for example, in the corners of the prismatic ones [80]. Additionally, cylindrical specimens are characterized by uniform deformation development along the entire perimeter of the specimen section. Thus, the edge effect problem is avoided in the tests.

The ratio between the diameter and the height of the concrete cylindrical specimen was 1/2 (15/30 cm). This ratio allows modeling of real full-scale compressed structural elements' behavior. Additionally, the chosen ratio guarantees the presence of a relatively large unrestrained area in the middle of the specimen's height [80], which is especially important for obtaining the descending branch. In this case, the element buckling problem is practically prevented. At the same time, due to some inaccuracy in the transmission of external compressive forces and concrete heterogeneity, additional stresses, associated with small (technological) eccentricities, may appear. This problem was solved by selecting a proper test setup.

4.3 Concrete Mixture Proportioning and Specimen Preparation

4.3.1 Concrete Mixture Proportioning

As it was mentioned above, the first research phase included two test series. Ready-mix concrete and concrete produced in laboratory conditions were used in the first and second test series, respectively. Concrete mixture for the first series was prepared using the following materials:

- locally produced composite 52.5N CEM II/AM-SLV Portland cement with portions of granulated blast furnace slag (S) and siliceous fly ash (V) from 18 to 30% [81];
- water-reducing and retarding admixture (G100X) with a density of 1.212 kg/dm^3;
- retarding and water-reducing admixture (HGP) with a density of 1.188 kg/dm^3;
- mix of three different types of locally available quarry sand with a fraction size of 0…1.2 mm, 0…2.5 mm, and 0…5 mm;

- locally available quarry crushed dolomite limestone with a coarse aggregate fraction of 9...16 mm according to [82].

The materials for preparing specimens for the second test series were:

- locally produced composite 42.5N CEM II/B-LL Portland cement [81] with portions of limestone between 21 and 35%;
- locally available quarry sand fraction of 0–4.75 mm;
- locally available quarry crushed dolomite limestone coarse aggregate fraction of 19/14 according to SI 3-1998 [82].

Basic properties of the concrete components are presented in Table 1. The final composition is given in Table 2.

Table 1 Concrete component properties.

Components	*Units*	*Values*
Cement density, ρ_{cem}	kg/m^3	2900
Standard cement consistency, K_{scc}	%	27.5
Sand fineness modulus, *FM*	–	1.85
Sand particle density, ρ_s	kg/m^3	2644.5
Crushed stone bulk density (oven-dry), ρ_{gb}	kg/m^3	1270.9
Crushed stone particle density, ρ_g	kg/m^3	2586.7

Table 2 Final concrete mixture proportioning for the first study phase.

Components	*Units*	*Values*
Cement, C	kg/m^3	448.4
Water, W	kg/m^3	224.2
Water-cement ratio, W/C	–	0.5
Sand, S	kg/m^3	602.2
Crushed stone, G	kg/m^3	1017.8
Fresh concrete density in compacted state, ρ_c	kg/m^3	2293

The final concrete mixture compositions for dry aggregates are presented in Table 3.

Concrete proportioning procedure for these two phases was based on the method of absolute volumes [83]. The methodology combines the following two principles of concrete mixture components' joint performance. First, it is well known that one of the main goals is to minimize the voids' number during mixture preparation since this parameter directly affects the final product strength. Second, the cement-sand mortar main function is to fill all voids. Combining the above

Table 3 Final concrete mixture proportioning for the second study phase.

Components	*Units*	*Series 1*	*Series 2*
Cement, C	kg/m^3	275	448.4
Water, W	kg/m^3	158	224.2
Water-cement ratio, W/C	–	0.57	0.50
Sand, S	kg/m^3	914	602.2
Crushed stone, G	kg/m^3	942	1017.8
Water-reducing and retarding admixture, WR	% of C	1.6	–
Retarding and water-reducing admixture, RW	% of C	0.1	–
Fresh concrete density in compacted state, ρ_c	kg/m^3	2293	2293

mentioned two ideas allows accurate calculation of concrete mixture proportioning, based on the initial water-cement ratio (W/C).

4.3.2 Specimen Preparation

In the first phase, the concrete mixture was cast into steel molds and compacted by tamping and rodding in 2 and 3 layers for cubic and cylindrical specimens, respectively [84, 85]. In the second research phase, internal and external vibrations were used to compact the fresh concrete in cylindrical and cubic molds, respectively [84].

Specimens in all test series were covered by a polyethylene film and stored for 24 h after casting [86] (Figure 4). After demolding, specimens were cured for 27 days in a fog room under a constant temperature of 23 ± 2°C and relative humidity of 95 ± 5% (Figure 5). The concrete mixture was prepared using a 0.045 m^3 rotating pan mixer when the calculated mixture volume was 0.043 m^3. Dried aggregates were used for all the mixtures. The compressive strength was obtained at 28 days, according to [1, 68] (Figure 6).

Fig. 4 Storing casted specimens covered by a polyethylene film for the first 24 h.

Fig. 5 Curing demolded specimens for 27 days in a fog room.

Fig. 6 Cubic specimen before compression.

4.4 The Absolute Volumes Method

The first step was calculating the initial water-cement ratio:

$$\frac{W}{C} = \frac{k\, f_{cem}}{f_{ck} + 0.5\,k\, f_{cem}} \qquad ...(1)$$

where f_{cem} is the cement class;

k is a coefficient depending on aggregate quality.

If the value of W/C is lower than 0.4, it should be recalculated with another value of the coefficient, *k*:

$$\frac{W}{C} = \frac{k\ f_{cem}}{f_{ck} - 0.5k\ f_{cem}} \qquad ...(2)$$

Further, water demand W_T was assigned according to concrete mix workability and coarse aggregate properties following [83]. The concrete mix workability was chosen according to [79]. After that 5 kg of dry coarse aggregate was used for obtaining the required fractional composition according to [87, 88] (Figure 7a). The aggregate was passed through a standard sieve set: 25 mm, 19 mm, 14 mm, 9.5 mm, 4.75 mm, 2.36 mm, 1.18 mm (Figure 7b). A special "W.S. Tyler RO-TAP RX-30-10" sieve shaker was used for automatic sieving (Figure 7c). After sieving the residue weight on each sieve was calculated in percent of the initial one.

(a) (b) (c)

Fig. 7 Sieving equipment: (a) 5 kg sample of dry coarse aggregate; (b) standard sieve set; (c) "W.S. Tyler RO-TAP RX-30-10" sieve shaker.

Then standard cement consistency was obtained following [89]. For this purpose, 500 g of cement and 125 g of water were initially used. The water quantity was further corrected in order to achieve the required standard consistency. The cement paste was mixed using a special mixer (Figure 8) according to the following mode: mixing 90 seconds, pause of 30 seconds, and additional 90 seconds of mixing.

Fig. 8 Cement mortar mixer “JJ-5”.

After mixing, the paste was put into the Vicat apparatus mold and placed on a base plate that was pre-greased by a chemical neutral oil. The percentage of water demand by weight of the dry cement determines its standard consistency. If the standard consistency is different from 27%, the water demand should be reduced by 3…5 kg per each percentage point below 27% or increased if it is above 27%.

After that the fine aggregate fineness modulus (FM) was obtained according to standard methodology following [87, 88]. At the first step, 2 kg of dry fine aggregate were prepared. The following step was sieving the prepared sample through the standard sieve set: 4.75 mm, 2.36 mm, 1.18 mm, 0.6 mm, 0.3 mm, and 0.15 mm. A similar automatic sieving method was used for the coarse aggregate. After sieving the residual material weight at each sieve was obtained and the ratio between each fraction and the initial weight was calculated.

Finally, the fineness modulus was obtained:

$$FM = \frac{A_{2.36} + A_{1.18} + A_{0.60} + A_{0.30} + A_{0.15}}{100} \quad ...(3)$$

where A_i is a cumulative residue of each fraction;

$$A_i = a_{2.36} + a_{1.18} + ... + a_i;$$

a_i is a part of material on a sieve, relative to the initial weight, %;

$$a_i = \frac{m_i}{m} \cdot 100\%;$$

m_i is the weight of material retained on sieve, kg;
m is the initial sand weight, kg.

If the fine aggregate fineness modulus is less (higher) than 2, then the water demand should be increased (decreased) by 3…5 kg for each 0.5 units below (above) 2.

At the following step, the cement consumption was calculated:

$$C = \frac{W_T}{W/C} \quad ...(4)$$

If the cement consumption exceeds 400 kg, water demand should be increased:

$$W = W_T + 0.1(C - 400) \quad ...(5)$$

Next, a coefficient of coarse aggregate grains expansion by a cement-sand mortar, α, was determined according to the calculated cement consumption and W/C.

The coarse aggregate consumption was calculated as follows:

$$G = \frac{1}{\left(\frac{1}{\rho_{gb}} - \frac{1}{\rho_g}\right)\alpha + \frac{1}{\rho_g}} \quad ...(6)$$

where ρ_{gb} is a loose (oven-dry) bulk density of coarse aggregate, kg/m^3 [90];

ρ_g is a coarse aggregate particle density, kg/m^3 [91].

The fine aggregate content was calculated as follows:

$$S = \left(1 - \frac{1}{\rho_{cem}} - \frac{1}{\rho_g} - \frac{1}{1000}\right)\rho_s \quad ...(7)$$

where ρ_s is the fine aggregate particle density, kg/m^3 [91].

Finally, a trial concrete mixture was prepared according to the calculated composition. The fresh concrete mixture consistency [92] and its real density after compacting [93] were determined. The composition was adjusted accordingly in order to obtain the required concrete mixture parameters.

4.5 Test Setup

4.5.1 First Phase

It is known that a displacement rate-controlled test should be carried out to obtain the appropriate descending branch of σ_c vs. ε_c relationship [94]. Thus, for achieving the aims of the first research phase a new test setup was proposed.

A servo-hydraulic closed-loop class 1 testing machine (with a maximum error of 1% FS (1% for full scale measurement)) and a maximum load of 3000 kN ("Controls C5600") was used to test concrete cylinders under axial compression [80]. It was controlled by a stand-alone multi-test computerized console "Controls 50-C20M82" (Figure 9). The tests were conducted under a displacement control mode at a constant loading rate of 0.01 mm/min. Control signal was obtained by a sensor measuring loading head displacement (Press SLDT) (Figure 10). The data obtained from this sensor was additionally used for the specimen post-peak behavior analysis that will be described below. An external pressure sensor was installed in the testing machine oil line to measure the applied load.

Fig. 9 Stand-alone multi-test computerized control console "Controls 50-C20M82".

The following measurement equipment was used to obtain the compressed concrete stress-strain relationship. Eight strain-gauges (SGs) with 30 and 60 mm bases were used in the specimen longitudinal direction

Fig. 10 Sensor measuring loading head displacement.

to investigate the behavior in the ascending branch. SG sensors were placed in pairs (one 30 mm and one 60 mm SG at 90 degrees intervals) to duplicate the obtained experimental results. Both sensor types had a resolution of up to 0.1 micro-strains, a resistance of 120 Ohm, and a gauge factor of 2.09 and 2.08 for 30 and 60 mm SGs, respectively.

Two circumferential 60 mm long SGs were used in the transverse direction. Additionally, two longitudinal highly accurate transducers (HAT) with a 50 mm measurement base, sensitivity of 0.02 microns, and a ± 1.5 mm strokes were attached to the specimen's surface in the transverse SGs location zone (Figure 11). Thus, 12 sensors were used for testing each specimen. All sensors were placed in the specimen central zone, which is considered to be uninfluenced by boundary conditions, i.e., friction forces between the testing machine plates and specimen surfaces.

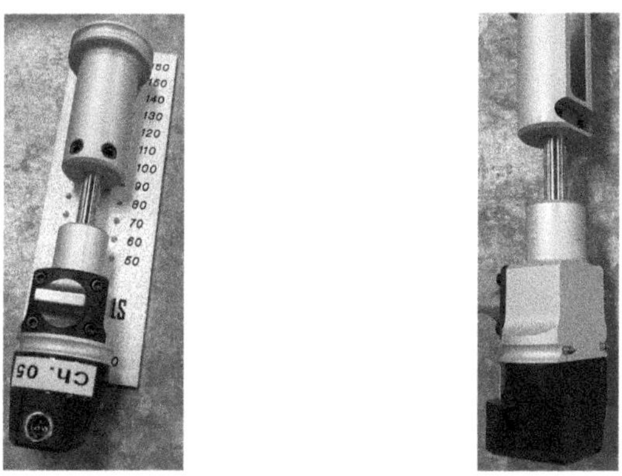

(a) (b)

Fig. 11 HAT device: (a) front view; (b) back view.

The test frame rigidity corresponded to normative requirements for excluding the influence of testing machine properties on post-peak concrete behavior records [95]. It is known that one of the predominant factors determining successful measurement of concrete $\sigma_c - \varepsilon_c$ relationship in the descending branch is joint performance of testing machine and compressed specimen. Thus, additional loading platens were designed and produced corresponding to the requirements provided in [96]: their thickness was 50 mm and diameter was 2 mm larger than the specimen's one. New mounts for linear displacement transducers were attached to these platens (Figure 12).

Fig. 12 Testing system with additional loading platens: 1–platens; 2–mounts for linear displacement transducers.

However, if the rigidity of stainless steel, used for the platens production, is not considered, the obtained stress-strain relationship shows fully elastic behavior of the specimen almost until failure and deformation values reach 5‰ at failure (Figure 13). This is explained

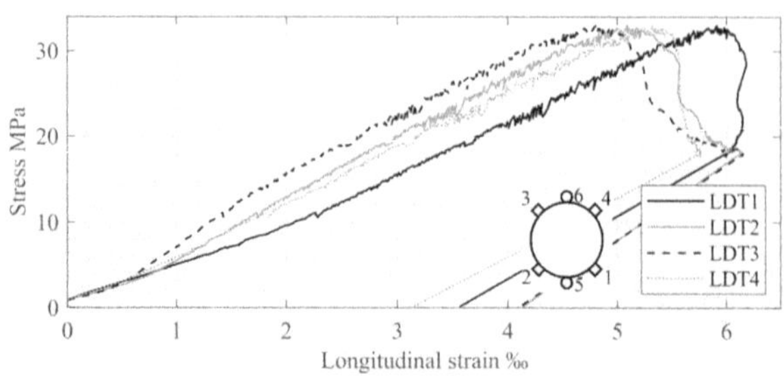

Fig. 13 Experimental dependencies σ_c vs. ε_c using LDTs and additional loading platens.

by the fact that the longitudinal LDTs measured the deformations of the concrete specimen and also those developing in the loading platens.

Finally, four LDT with a 50 mm stroke and a high resolution (<1 μm) were mounted between the testing machine original loading plates in order to obtain accurate response of a specimen in the post-peak stage (descending branch). A general view of the test final setup version is shown in Figure 14.

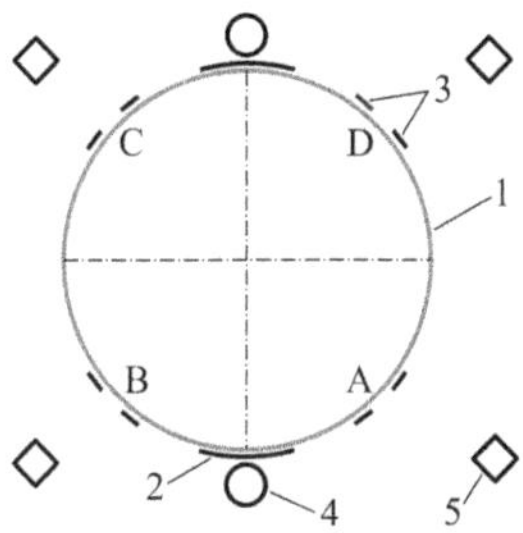

(a)

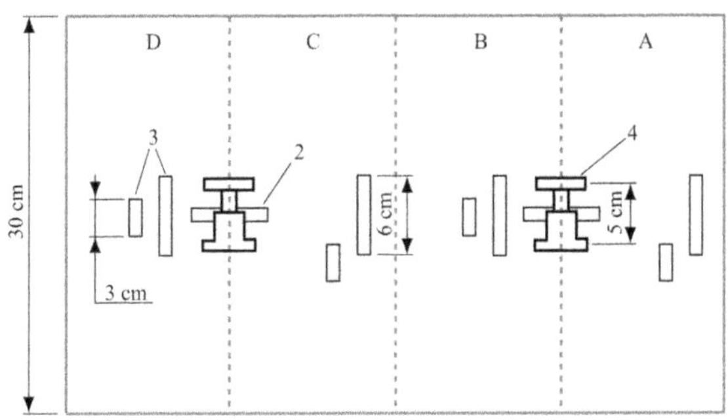

(b)

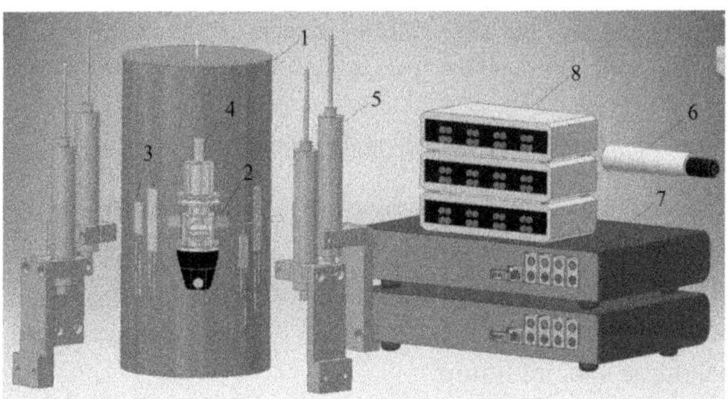

(c)

Figure 14 (*continued*).

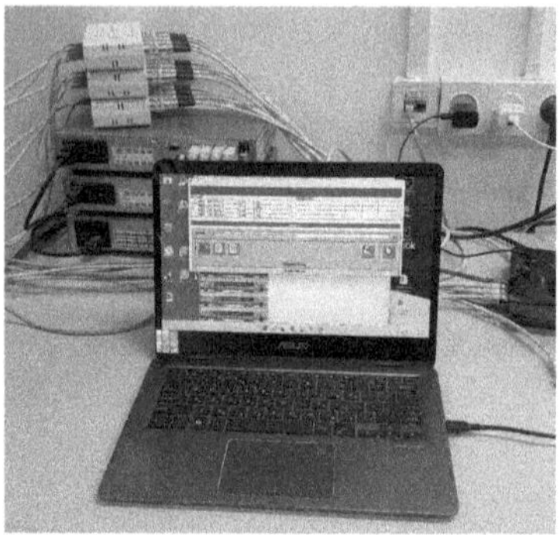

(d)

(e)

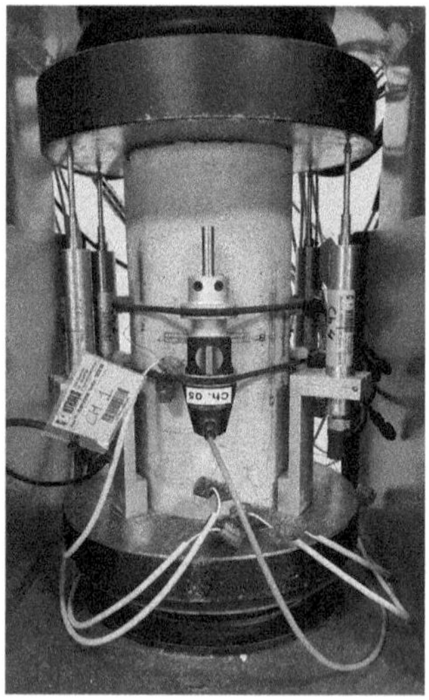

(f)

Fig. 14 Final test setup: (a) measurement device location scheme; (b) specimen surface development; (c) general view; (d) measuring equipment; (e) specimen before loading; (f) specimen under loading.

4.5.2 Test System Preparation Methodology

A special stencil was proposed to be used in order to simplify the device's positioning on a tested specimen (Figure 15). A surface without visible voids was selected to place the measurement devices. After that, the surface at the sensor places was cleaned with fine sandpaper and dust was removed using a special liquid.

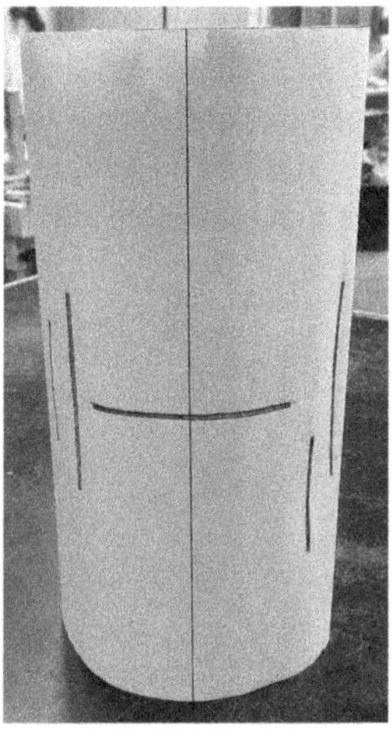

Fig. 15 Marking SG sensor's position using the stencil.

A technique for the SGs installation, developed by Micro-Measurements and improved by the authors, was used at this phase. The first step after cleaning and dedusting the installation zone was filling the pores using a thin epoxy layer, further abraded after 24 hours curing. After additional cleaning and dedusting using various conditioners and neutralizers, a special gluing procedure, requiring high accuracy and speed of execution, was used. After applying the adhesive, the sensor should be glued within 3–5 seconds, while being aligned according to the previously applied marks (Figure 16).

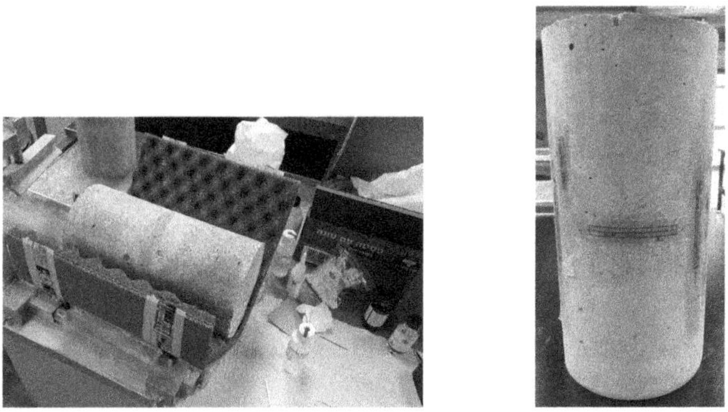

(a) (b)

Fig. 16 Attaching SG sensors: (a) specimen prepared for sensor installation; (b) specimen with attached sensors.

Next, the wires were soldered to plates with connecting terminals (Figure 17). Finally, the sensor resistance was checked by a tester. The upper surface of the cylindrical specimens was leveled by a thin layer of sifted fine grain sand. It allowed an additional effect of reducing the friction between the testing machine plate and a specimen.

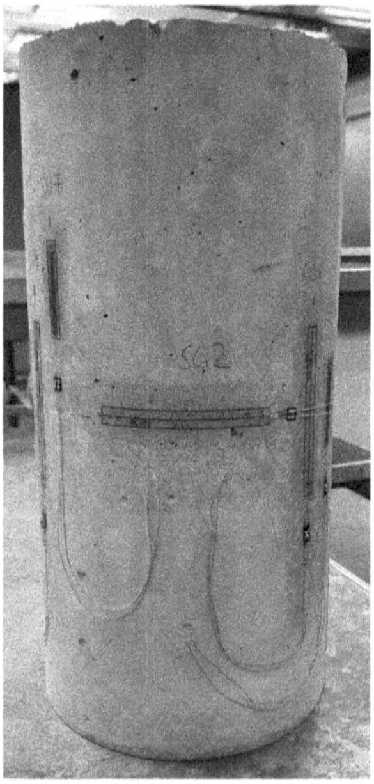

Fig. 17 Specimen prepared for testing.

4.5.3 Second Phase

The measurement system was selected according to the following specimen features that should be obtained:

- loading force;
- longitudinal deformations;
- transverse deformations.

To carry out the axial compression test, the same testing machine as for the first phase was used. However, it was controlled by a stand-alone automatic control console "Controls Sercomp 7" (Figure 18). This console allows setting the loading rate both in automatic and manual control modes. The applied load was measured by an external pressure sensor, which was installed in the testing machine oil line. The maximum error during calibration was 0.01% FS and the full-scale range was equal to 1800 kN. Additionally, the testing machine accuracy was determined by the minimum scale range, which was equal to 0.1 kN. Thus, the maximum permissible error can be 30 kN for a full-scale range, which corresponds to 0.4 MPa stress in a cylindrical specimen.

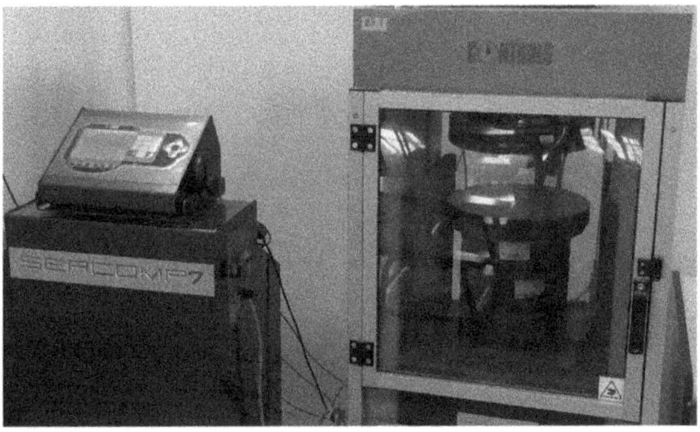

Fig. 18 "Controls C5600" testing machine with stand-alone automatic control console "Controls Sercomp 7".

For selecting sensors, measuring longitudinal and transverse deformations in the provided specimen type, the following preliminary calculation was made. At the beginning the initial stress in the loading process was selected as equal to 0.5 MPa corresponding to the requirement for measuring the modulus of elasticity [67, 97]. Corresponding initial longitudinal strain for the concrete class C25/30 is:

$$\varepsilon_{long\ init} = \frac{\sigma_{init}}{E_c} = \frac{0.5 \cdot 10^6}{30 \cdot 10^9} = 0.017\ ‰ \qquad ...(8)$$

Then the longitudinal shortening for the 150 mm measurement base was calculated in the following form:

$$\Delta l_{init} = \frac{150 \cdot 0.017‰}{1000‰} = 0.0034\ \text{mm} \qquad ...(9)$$

Therefore, transducer measurement accuracy equal to 1 μm is sufficient. Considering sensors for transverse deformation measuring, the accuracy of 0.2 μm is sufficient, as the average value of Poisson's ratio for NSC equals 0.2 [2]. As a result, the measuring system described below was chosen.

Four linear displacement transducers (LDT) with a 50-mm stroke and a high-measure resolution (<1 μm) were used to measure the longitudinal deformations (Figure 19). The sensors were located in the specimen's central zone in order to exclude the influence of the edge effect associated with the friction forces, occurring at the specimen contact area between the testing machine plates. The base of the longitudinal deformation measurement was 15 cm. The upper view of the tested specimen and the measurement devices is shown in Figure 20. Figure 21 presents the device's location on the specimen's surface development.

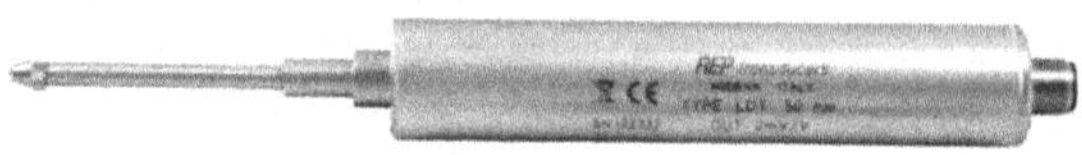

Fig. 19 LDT device.

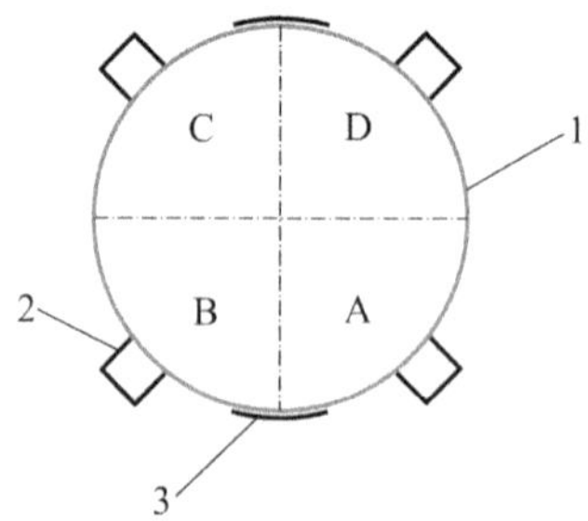

Fig. 20 Measurement device position scheme: 1. Cylindrical concrete specimen; 2. LDT devices; 3. Strain-gauge (SG) sensors; A, B, C, and D – specimen sectors.

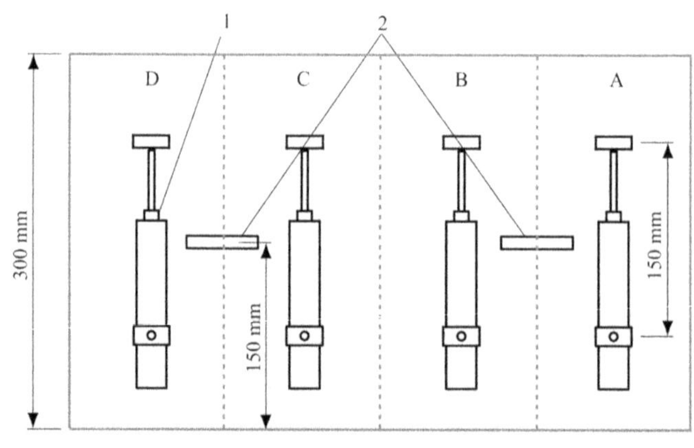

Fig. 21 Specimen surface development: 1. LDT devices; 2. SG sensors; A, B, C, and D – specimen sectors (according to Figure 20).

A technological eccentricity issue, caused by inaccuracy in the applied force transition, has occurred in this test phase. This problem was solved by using four sensors located on the tested specimen opposite sides (see Figure 20). Such sensor distribution allowed obtaining more accurate data that revealed the presence of eccentricity in stress distribution. The LDT sensors were attached by originally developed mounts, allowing independent deformation transfer to each sensor (Figure 22).

Fig. 22 LDT device mount.

Two strain-gauges with a 60 mm base, a measure resolution up to 0.1 micro-strains, a resistance of 120 Ohms, and a gauge factor of 2.08 were used to measure the transverse deformations (Figure 23). The SG length was chosen based on the requirement of exceeding the coarse aggregate maximum dimensions by at least three times [89].

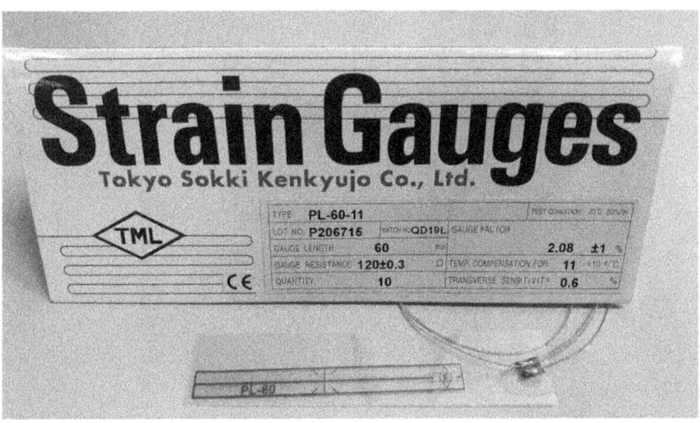

Fig. 23 Strain-gauge.

The measurement equipment selection includes not only the sensors, but also the data recording devices, namely the data logger and Wheatstone bridge. A Wheatstone bridge is required to connect SG sensors in a predetermined sequence with a specific resistance algorithm. A data logger is required for measuring and processing the data from all sensors with subsequent transferring them to a PC in a digital form. Both chosen items were also produced by Controls. The general view of the test setup is presented in Figure 24 and the specimen prepared for the test is shown in Figure 25.

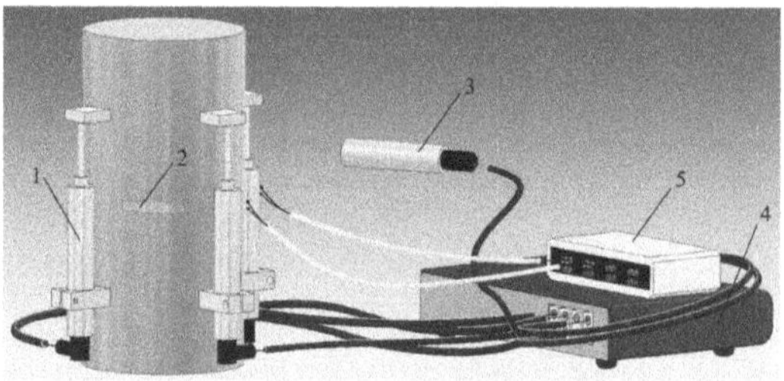

Fig. 24 Measurement devices: 1. LDT devices; 2. SG sensors; 3. Load sensor; 4. Data logger; 5. Wheatstone bridge.

Fig. 25 Specimen prepared for testing.

4.5.4 Testing System Preparation Methodology

As it was mentioned above, a new stencil was used to simplify the devices' positioning on a tested specimen surface (Figure 26). Like in the previous phase, a surface without visible voids was selected to place the measurement devices. After that, the surface at the mount places was cleaned with fine sandpaper and dust was removed using a special liquid.

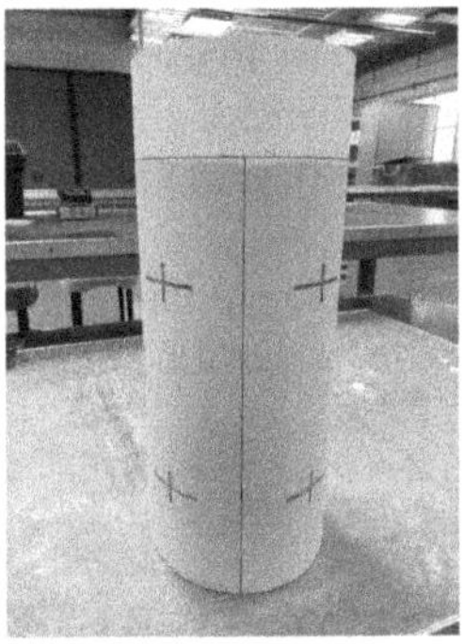

Fig. 26 Marking the mount positions using the stencil.

A special epoxy adhesive with a proper ability was selected to glue the mounts to the concrete specimen. It should be mentioned that a thin layer of glue should be used to avoid the glue elastic deformations during a test. The mount should be leveled to maintain vertical and horizontal directions (Figure 27).

Fig. 27 Mount positioning.

The SG sensors installation zone, corresponding to the sensor length with its wires, equal to 95 mm, was cleaned and dedusted. The process of surface preparation for attaching SG sensors was described in detail in Section 4.5.2. After that, the sensor resistance with the already soldered plate including connecting terminals and wires was checked by a tester. The resistance should be equal to 120 ± 3 Ohm. Next, the sensor was placed on a special paper and covered with Loctite AA330 glue (Figure 28a). Loctite AA7388 activator was then sprayed onto the specimen adhesion area and the sensor was adhered to the specimen, followed by firm pressure, smoothing and flattening in the horizontal plane (Figures 28 b and c). The plate with connecting terminals should also be glued, because its function is to remove the load from the wires connected to the sensor (Figure 28d). According to the Loctite A330 user manual, it is necessary to allow the glue to harden for 1 hour, after which the specimen can be tested.

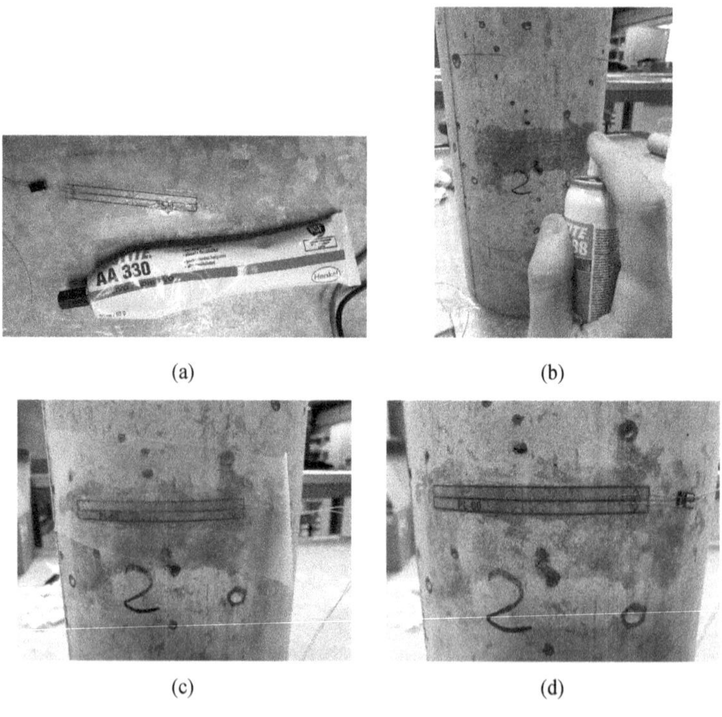

(a) (b) (c) (d)

Fig. 28 SG sensors glueing: (a) applying adhesive to a sensor; (b) activator application in the glueing area; (c) sensor glueing; (d) glued sensor including the plate with connecting terminals.

A thin layer of sifted fine grain sand was also used to level the upper surface of the cylindrical specimens (Figure 29).

(a) (b)

Fig. 29 Specimen with fine grain sand on the upper surface: (a) general view; (b) sand leveling process.

5

Experimental and Numerical Verification of the Theoretical Stress-Strain Model for Compressed Concrete

5.1 Investigation of Compressed Concrete Stress-Strain Relationship

Experimental investigation of the concrete stress-strain relationship can be divided into two stages: the ascending branch and the descending one. The ascending branch can be obtained both by load and displacement rate control testing and it was widely investigated [98]. It should be mentioned that obtaining the descending branch of the compressed concrete stress-strain diagram is also an important step, because this branch, together with the ascending one, determines the potential of compressed concrete, which is essential for proper design of concrete structures subjected to dynamic loads [3].

It should be noted that, when analyzing concrete response in the descending branch, some researchers considered that the modulus of elasticity is negative [99]. But in the authors' opinion, the exact stage up to which the stress reduction occurs is important. A definite answer to this question is given in RILEM recommendations [96], stating that the concrete stress decreases to half of the load-bearing capacity ($0.5f_c$). The present study is also focused on the experimental, numerical, and theoretical investigation of this problem.

The question is if the descending branch should be considered as a concrete property on a material level (but not on structural level), or whether it is a result of deformations' development caused by experimental techniques [100, 101, 102]. A corresponding theory was developed by Feretti. According to this theory, obtaining stresses by dividing the applied load by the initial specimen cross-sectional area is incorrect after reaching their peak value [103, 104, 105]. However, the authors do not agree with this approach, therefore the specimen initial cross section and potential energy were investigated. Consequently, according to this theory the potential energy will grow after passing the limit stress value (at the descending branch). An example of the descending branch presence in real structure behavior is a failure scheme of a shell with a dent [106].

Another issue related to the descending branch investigation is considering the joint work of a testing machine and a compressed specimen. The influence of a testing machine's insufficient stiffness on the obtained deformations in the compressed concrete specimen was identified [52, 107]. Cylindrical normal strength concrete specimens with 10 cm diameter and 25 cm height were tested by axial loading [98]. Cylindrical specimens with 10 cm diameter and 20 cm height and prismatic ones with 20 cm height and 10 cm side length made of normal and high strength concrete were investigated [108]. The obtained results confirmed the influence of insufficient testing machine stiffness on the recorded strains in compressed concrete. Therefore, using a system, allowing correction of the obtained data by compensating the released energy in the post-peak concrete behavior stage, was proposed [52, 98]. Additionally, a method, determining a correction factor that considers different stiffness of testing machines by measuring modulus of elasticity from deformations in a specimen and between testing machine plates, was developed [98].

As a result, using a closed-loop testing machine, controlled by specimen strain feedback, was proposed for obtaining a full stress-strain diagram for compressed concrete, including the descending branch [96, 108, 109]. Some other requirements for test specimens were formulated based on available experimental results [80, 96]:

- providing a certain ratio of specimen height to its cross-section dimension to ensure the presence of a central area in the specimen height, which does not depend on the boundary conditions. This ratio also affects the descending branch shape: its slope increases as the ratio becomes higher, which means a more brittle specimen failure;
- taking into account the friction between a specimen and the testing machine plates and reduce this influence as far as possible;
- minimum three longitudinal external sensors should be placed around a specimen between the testing machine plates to obtain the entire deformation form;
- the loading rate should correspond to static (but not dynamic) loading (about 0.1 μm/s).

The present research phase was devoted to in-depth experimental investigation of the above-mentioned problems, including developing new methods for their implementation. In particular, this investigation was based on the previous theoretical results and on generalizing the stress-strain model [3], as well as the Mini max Principle [58], the Quasi-isotropic concept of the RC element at ULS [59], and the Structural Phenomenon [60].

5.2 Experimental Data Processing Methodology

Experimental data processing included three stages. At the first stage, data obtained from the longitudinal SGs and from the HAT sensors were analyzed. These data allowed obtaining the ascending branch of the compressed concrete stress-strain diagram. At the second stage, data from sensors, attached to the testing machine (PressLDT and LDTs), were evaluated (see Figure 14a [Chapter 4]). These data were used to obtain the descending branch of the stress-strain diagram. At the third stage, the curves, obtained in the first and second stages, were combined into a complete stress-strain relationship of compressed concrete specimens.

5.3 Experimental Results and Stress-Strain Diagrams of Compressed Concrete

The obtained experimental values for the main stress-strain ($\sigma_c - \varepsilon_c$) diagram parameters as well as relations between strains, strength, and stresses from elastic to ultimate stages are shown in Tables 1 and 2.

Typical failure schemes of the tested specimens are presented in Figure 1, which shows that the failure was accompanied by the vertical cracks' development process, which is further described in detail (see Section 6.3).

The behavior of compressed specimens is determined by specific points, corresponding to maximum elastic stresses and deformations (point 1 in Figure 2), specimen elastic potential (point 2 in the Figure 2), ultimate limit state (point 3 in the Figure 2) and failure on the descending branch level (point 4 in the Figure 2). The specified values at point 1 are: $\varepsilon_{c\ el}$ = 0.5‰ and $\sigma_{c\ el} = 0.5f_c$. At point 2 the values correspond to $\varepsilon_c = 2\varepsilon_{c\ el}$ = 1.0‰ and $\sigma_c = f_c$. At point 3 the values correspond to ε_c = 2.0‰ and $\sigma_c = f_c$, and at point 4 ε_c = 3.5‰ and $\sigma_c = 0.5f_c$.

Table 1 Experimental values of the $\sigma_c - \varepsilon_c$ diagram ascending branch parameters.

Specimen	$\varepsilon_{c\,el}$, ‰	$\varepsilon_{c\,el\,ul}$, ‰	ε_{c1}, ‰	$\frac{\varepsilon_{c\,el}}{\varepsilon_{c\,el\,ul}}$	$\sigma_{c\,el}$, *MPa*	f_c, *MPa*	$\frac{\sigma_{c\,el}}{f_c}$
Sp1.1	0.50	1.03	2.00	0.49	17.50	36.50	0.48
Sp1.2	0.50	1.06	2.16	0.47	18.10	38.70	0.47
Sp1.3	0.50	1.03	2.03	0.49	20.99	41.60	0.50
Average	0.50	1.04	2.06	0.48	18.86	38.93	0.48
Sp2.1	0.50	1.12	1.72	0.45	11.92	24.94	0.48
Sp2.2	0.50	1.06	1.80	0.47	11.35	22.35	0.51
Sp2.3	0.50	1.19	1.79	0.42	11.86	26.40	0.45
Average	0.50	1.12	1.77	0.45	11.71	24.56	0.48

Table 2 Experimental values of the $\sigma_c - \varepsilon_c$ diagram descending branch parameters.

Specimen	$\varepsilon_{c\,ul}$, ‰	$\sigma_{c\,ul}$, *MPa*	$\frac{\sigma_{c\,ul}}{f_c}$
Sp1.1	3.50	24.10	0.66
Sp1.2	3.50	22.40	0.58
Sp1.3	3.35	27.53	0.66
Average	3.45	24.67	0.63
Sp2.1	3.12	14.91	0.60
Sp2.2	3.19	11.27	0.50
Sp2.3	3.00	16.05	0.61
Average	3.10	14.08	0.57

Fig. 1 Typical specimen failure schemes.

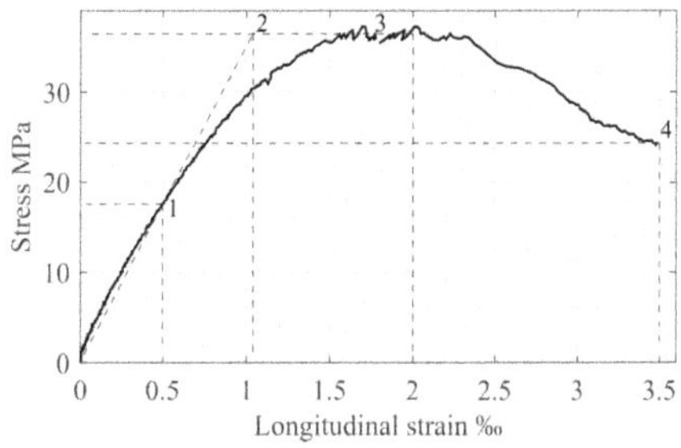

(a)

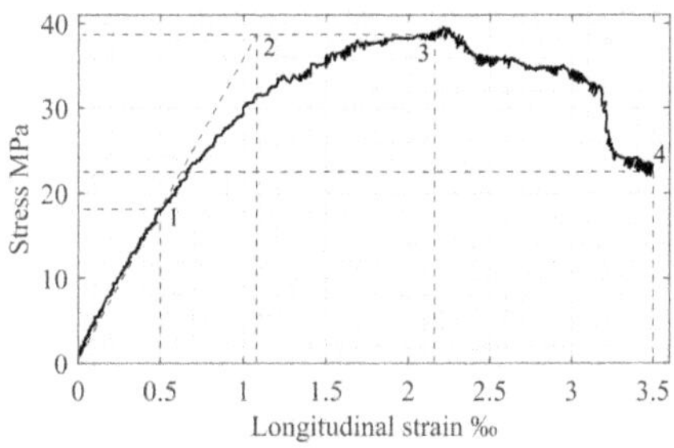

(b)

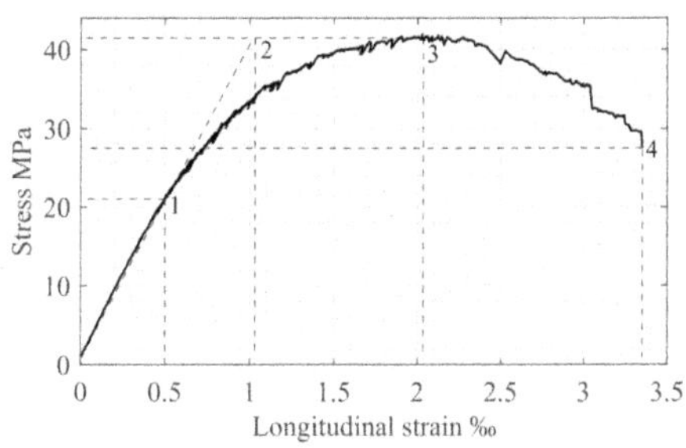

(c)

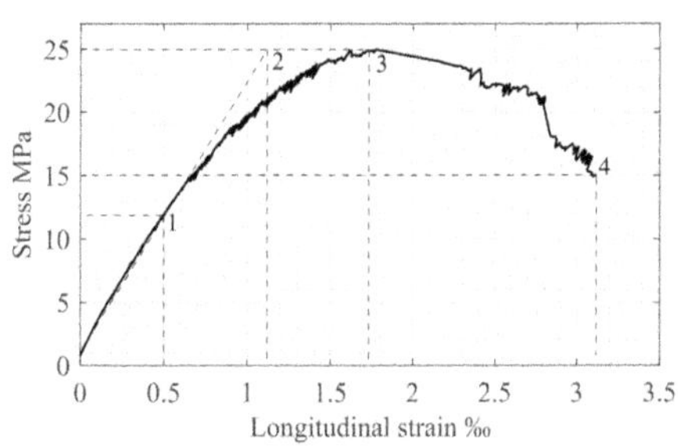

(d)

Figure 2 (*continued*).

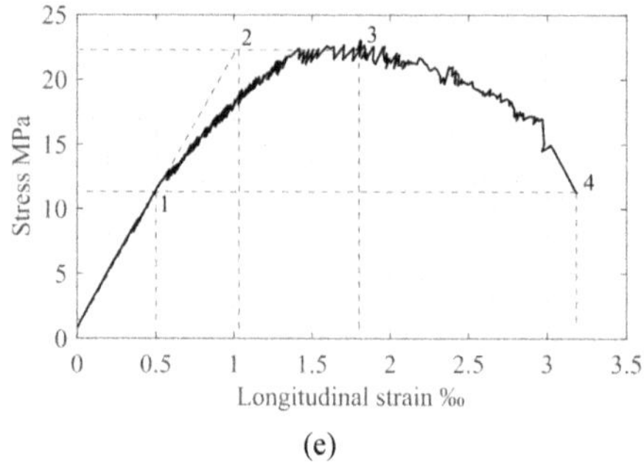

(e)

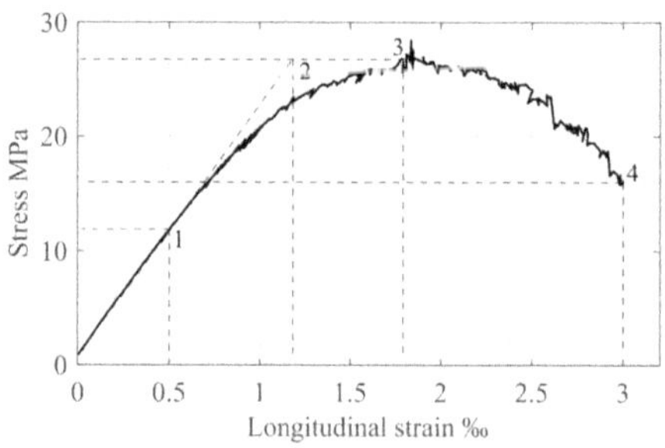

(f)

Fig. 2 Experimental relationships $\sigma_c - \varepsilon_c$: (a) specimen Sp1.1; (b) specimen Sp1.2; (c) specimen Sp1.3; (d) specimen Sp2.1; (e) specimen Sp2.2; (f) specimen Sp2.3; point 1 in each plot corresponds to the maximum elastic stresses and deformations; point 2 to the specimen elastic potential; point 3 to the ultimate limit state, and point 4 to the failure on the descending branch level.

5.4 Comparison of Experimental and Theoretical Results

Analysis of variation coefficients (VC) based on standard deviations for the experimental data presented in Tables 1 and 2 shows sufficient convergence of the obtained results (Table 3). The maximum VC value is 14.5%, which is acceptable for concrete, considering its natural features (e.g., material heterogeneity) as well as possible load-application eccentricities, etc.

Table 3 Variation coefficients for the $\sigma_c - \varepsilon_c$ diagram parameters.

	Test Series	$\varepsilon_{c\,el}$	$\varepsilon_{c\,el\,ul}$	ε_{c1}	$\frac{\varepsilon_{c\,el}}{\varepsilon_{c\,el\,ul}}$	$\frac{\sigma_{c\,el}}{f_c}$	$\varepsilon_{c\,ul}$	$\sigma_{c\,ul}$	$\frac{\sigma_{c\,ul}}{f_c}$
VC, %	1	0.00	1.36	3.37	1.35	3.18	2.05	8.65	6.12
	2	0.00	4.73	2.01	4.72	5.00	2.53	14.50	8.21

As mentioned above, the main parameters for concrete behavior under compressive load are stresses and strains, determined by points 1–4 (see Figures 20 [Chapter 2] and 2 in this chapter). A comparison of the corresponding experimental and theoretical data is presented in Table 4. Consequently, a graphical representation of theoretical and experimental $\sigma_c - \varepsilon_c$ diagrams is shown in Figure 3. The results of comparing the theoretical and experimental potentials values of the $\sigma_c - \varepsilon_c$ diagrams [3] are presented in Table 5.

Table 4 Comparison of average experimental and theoretical values of the main $\sigma_c - \varepsilon_c$ diagram parameters.

Data		$\varepsilon_{c\,el}$	$\varepsilon_{c\,el\,ul}$	ε_{c1}	$\frac{\varepsilon_{c\,el}}{\varepsilon_{c\,el\,ul}}$	$\frac{\sigma_{c\,el}}{f_c}$	$\varepsilon_{c\,ul}$	$\frac{\sigma_{c\,ul}}{f_c}$
Theoretical		0.50	1.00	2.00	0.50	0.50	3.50	0.50
Average Experimental	Series 1	0.50	1.04	2.06	0.48	0.48	3.45	0.63
	Series 2	0.50	1.12	1.77	0.45	0.48	3.10	0.57

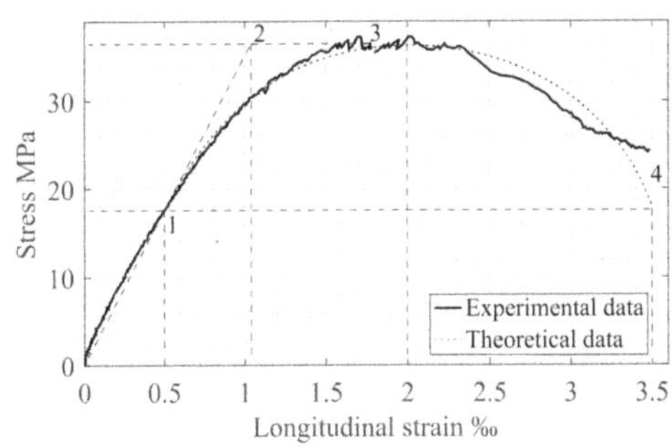

(a)

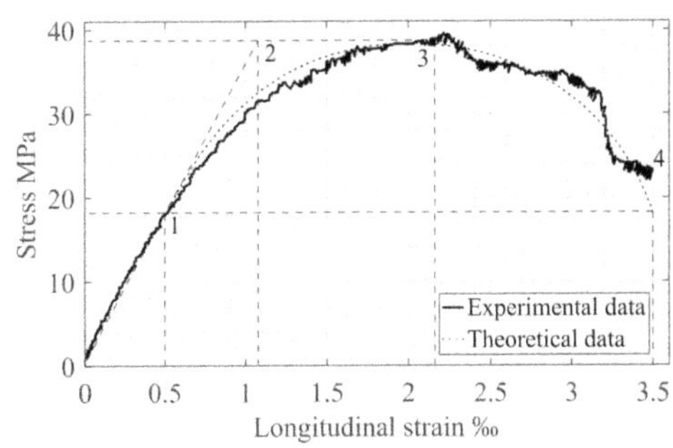

(b)

Figure 3 (*continued*).

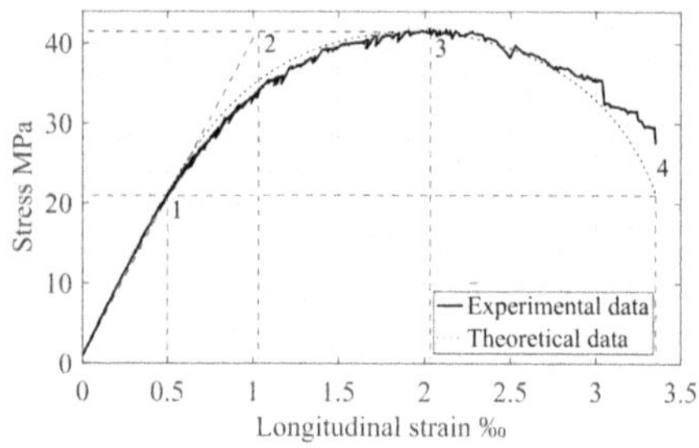

(c)

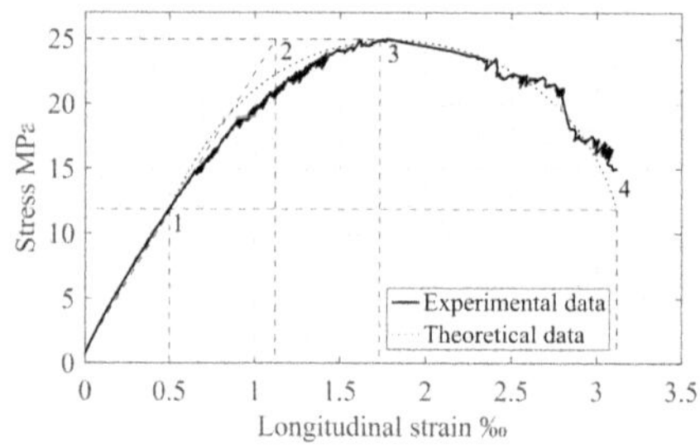

(d)

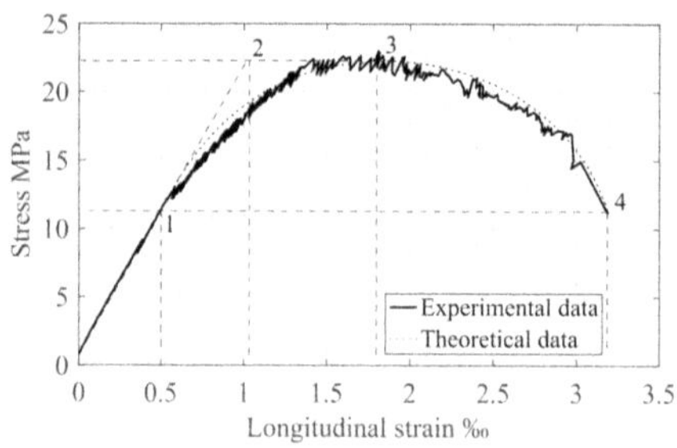

(e)

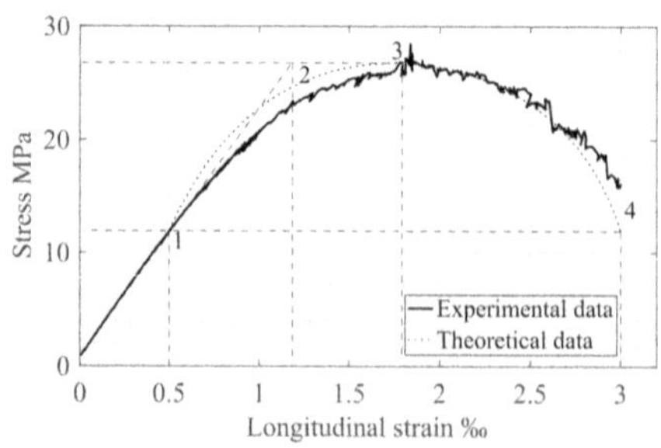

(f)

Fig. 3 Comparison of experimental and theoretical $\sigma_c - \varepsilon_c$ relationships: (a) specimen Sp1.1; (b) specimen Sp1.2; (c) specimen Sp1.3; (d) specimen Sp2.1; (e) specimen Sp2.2; (f) specimen Sp2.3.

Table 5 Experimental and theoretical values of elastic and plastic potentials and the relation between them.

Specimen	U_{el}^{Theor}, *kPa*	U_{el}^{Exper}, *kPa*	U_{pl}^{Theor}, *kPa*	U_{pl}^{Exper}, *kPa*	$\frac{U_{el}^{Exper}}{U_{pl}^{Exper}}$	*Average* $\frac{U_{el}^{Exper}}{U_{pl}^{Exper}}$
Sp1.1	31.94	32.87	63.88	65.81	0.50	0.53
Sp1.2	33.86	38.72	67.73	63.13	0.61	
Sp1.3	36.40	36.69	72.80	71.95	0.51	
Sp2.1	21.82	18.87	43.65	39.54	0.48	
Sp2.2	19.56	17.14	39.11	37.54	0.46	
Sp2.3	23.10	21.72	46.20	37.07	0.59	

Analysis of the data presented in Table 4 shows a rather good convergence. A comparison of elastic and plastic potentials, presented in Table 5, also demonstrates good agreement between theoretical and experimental results. Further comparison of the relation between U_{el}^{Exper} and U_{pl}^{Exper} confirms the Structural Phenomenon concept [60]: the elastic parameters increase twice when specimens reach the plastic stage. Taking these results into account, the entire range of the obtained experimental data and its comparison with the theoretical one completely confirms the proposed theoretical stress-strain model for compressed concrete.

5.5 Numerical Analysis of the Experimental Stress-Strain Relationships

Stresses and strains values as well as their dependencies, provided in the previous study phase, were used for numerical analysis. A comparison of the corresponding experimental and numerical data, using $\sigma_c - \varepsilon_c$ diagrams, is presented in Figure 4. The results of comparing the theoretical and numerical potentials of the stress-strain diagrams [3] are presented in Table 6.

The graphical representations in Figure 4 demonstrate high correspondence between the experimental and numerical $\sigma_c - \varepsilon_c$ relationships. Relative error *D* presented in Table 6 also shows rather good convergence between experimental and numerical data for the elastic potential, describing the elastic stage of concrete behavior. Considering the non-elastic stage, the maximum relative error reaches 17%, which is acceptable for such a complex material as concrete. Comparison of the relation between U_{el}^{Num} and U_{pl}^{Num}, obtained from experimental results, also confirms the Structural Phenomenon concept [60].

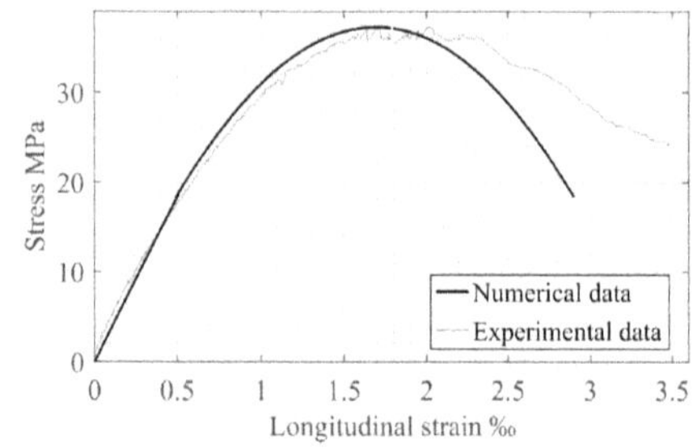

(a)

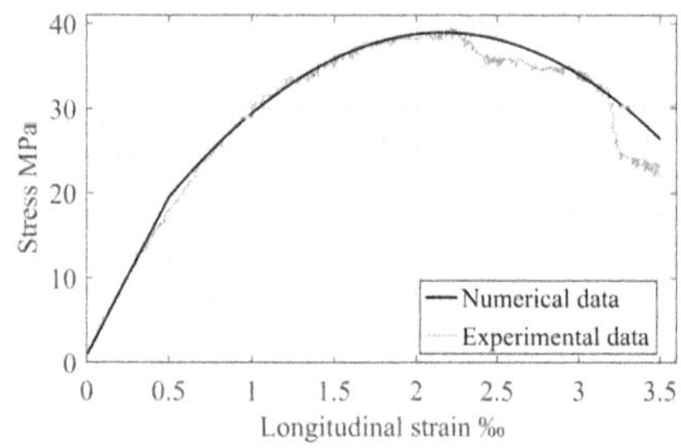

(b)

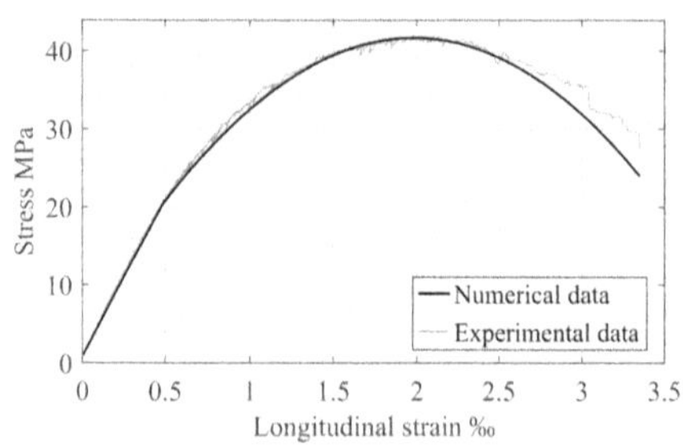

(c)

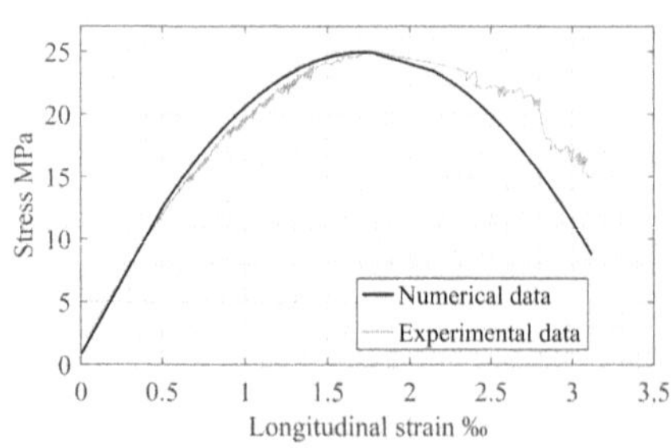

(d)

Figure 4 (*continued*)

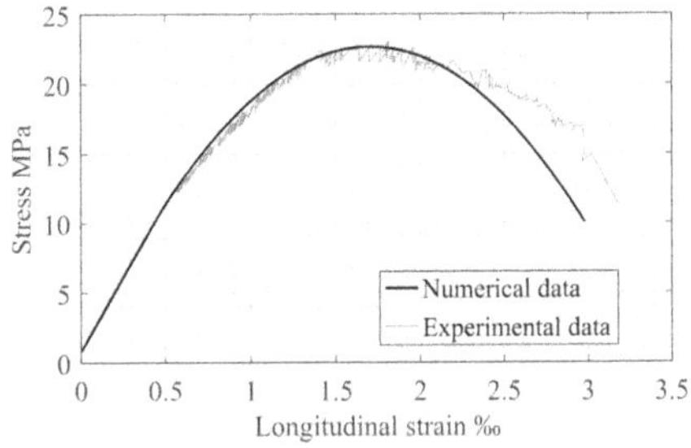

(e)

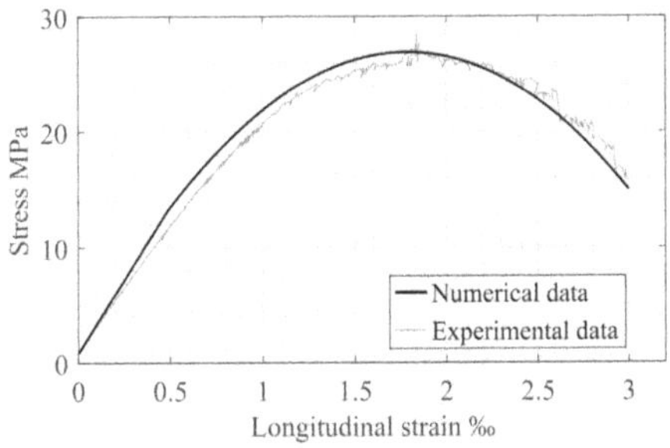

(f)

Fig. 4 Comparison of experimental and numerical $\sigma_c - \varepsilon_c$ relationships: (a) specimen Sp1.1; (b) specimen Sp1.2; (c) specimen Sp1.3; (d) specimen Sp2.1; (e) specimen Sp2.2; (f) specimen Sp2.3.

Table 6 Experimental and numerical values of elastic and plastic potentials and the relation between them.

Specimen	U_{el}^{Num}, *kPa*	U_{el}^{Exper}, *kPa*	*D*, %	U_{pl}^{Num}, *kPa*	U_{pl}^{Exper}, *kPa*	*D*, %	$\frac{U_{el}^{Num}}{U_{pl}^{Num}}$	*Average* $\frac{U_{el}^{Num}}{U_{pl}^{Num}}$
Sp1.1	30.86	32.87	6.13	54.58	65.81	17.06	0.52	0.51
Sp1.2	37.27	38.72	3.74	75.62	63.13	16.52	0.49	
Sp1.3	36.09	36.69	1.64	72.07	71.95	0.17	0.50	
Sp2.1	18.38	18.87	2.60	35.62	39.54	9.91	0.52	
Sp2.2	16.50	17.14	3.73	31.90	37.54	15.02	0.52	
Sp2.3	20.68	21.72	4.79	40.41	37.07	8.27	0.51	

Finally, all three stages, including the theoretical, experimental, and numerical investigations, are presented in Figure 5. Thus, full confirmation of the theoretical stress-strain model for compressed concrete is obtained.

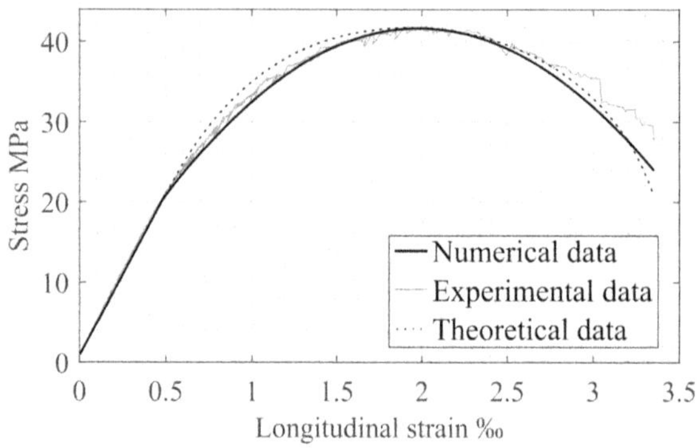

Fig. 5 Comparison of theoretical, experimental, and numerical $\sigma_c - \varepsilon_c$ relationships

6

Experimental Investigation of Concrete Transverse Deformations

6.1 Evaluation of Concrete Parameters in Elastic State

The experimentally obtained dependences between the longitudinal deformations and stresses in uniaxially loaded cylindrical concrete specimens, which were tested in the first test series, are presented in Figure 1. As it is evident from this figure, the dependencies are close to the linear and allow for the disclosure of concrete ultimate elastic deformations and modulus of elasticity.

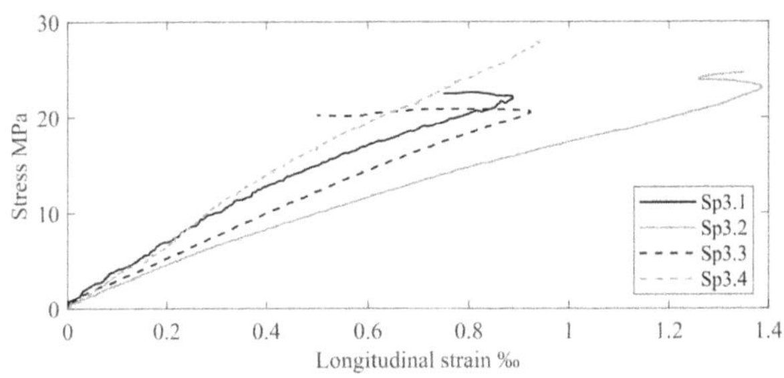

Fig. 1 Experimental dependencies σ_c vs. ε_c, obtained in the first test series.

Dependencies of transverse deformations vs. stresses are shown in Figure 2. These dependencies enable us to obtain the concrete Poisson's ratio as well as the pattern of inelastic concrete deformation development (which cannot be done using only longitudinal deformations). It should be mentioned that this possibility becomes available when the transverse deformations exceed the Poisson's ratio validity range, i.e., after the initiation of longitudinal cracks, when $\varepsilon_{trans} > \varepsilon_{ct\ ul}$.

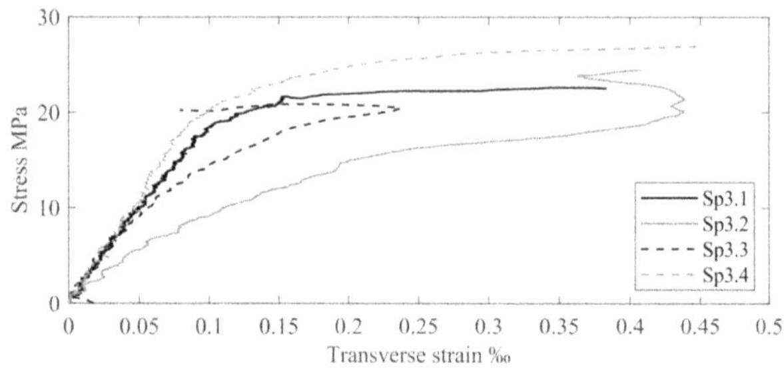

Fig. 2 Experimental dependencies σ_c vs. ε_{trans}, corresponding to the first test series.

Following Figure 2, the tested specimens exhibited elastic behavior in transverse direction up to about 70% of the concrete strength value (for the investigated concrete class the strength value ranges from 20.9 to 28.2 MPa). For example, the concrete strength of Sp3.1 is equal to 24 MPa, and the strength value, corresponding to the elastic stage, reaches 17 MPa, which is about 70% of the specimen strength. Further, transverse deformations grow non-linearly and indicate the inelastic deformations development that are not evident in the longitudinal direction (see Figure 1). The experimental results are summarized in Table 1.

Table 1 Experimental values of maximum elastic and ultimate transverse concrete strains, $\varepsilon_{c\ el}$, $\varepsilon_{ct\ ul}$, modulus of elasticity, E_c, and Poisson's ratio, v

Specimen	$\varepsilon_{c\ el}$, ‰	$\varepsilon_{ct\ ul}$, ‰	E_c, *MPa*	v
Sp3.1	0.678	0.100	27700	0.15
Sp3.2	0.453	0.100	27600	0.22
Sp3.3	0.556	0.086	26740	0.16
Sp3.4	0.605	0.092	32000	0.15
Average	0.573	0.095	28510	0.17

The obtained experimental data correspond to those available in *Eurocode 2* [1]. A comparison of the known Poisson's ratios data is presented in Table 2. As evident from this table, the Poisson's ratios for all of the tested specimens are close to the known results.

Table 2 Comparison of Poisson's ratios.

References				
SI 466 [2]	*Bondarenko and Suvorkin [110]*	*BR 52-101 [111]*	*Eurocode 2 [1]*	*Iskhakov [62]*
0.15 … 0.25	0.20	0.20	0.20	0.15 ... 0.20

It should be also noted that according to the previously proposed theoretical model [3], the maximum elastic longitudinal deformations are within 0.5‰, and the ultimate ones are equal to 1‰, which is also confirmed by the obtained experimental data.

The data obtained in the second test series show similar tendencies in the specimens' behavior and confirm the previously presented results (Figures 3 and 4).

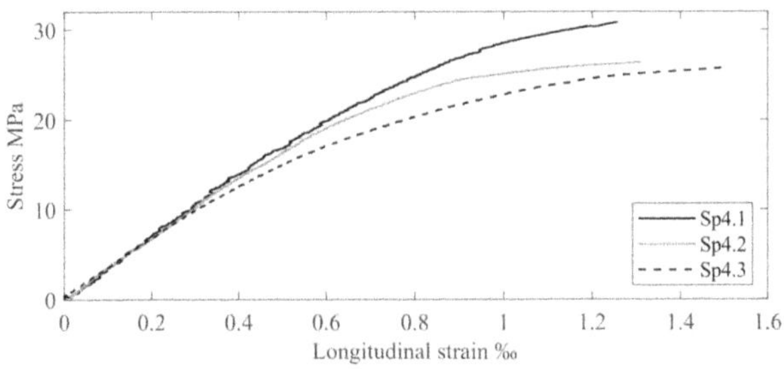

Fig. 3 Experimental dependencies σ_c vs. ε_c, obtained in the second test series.

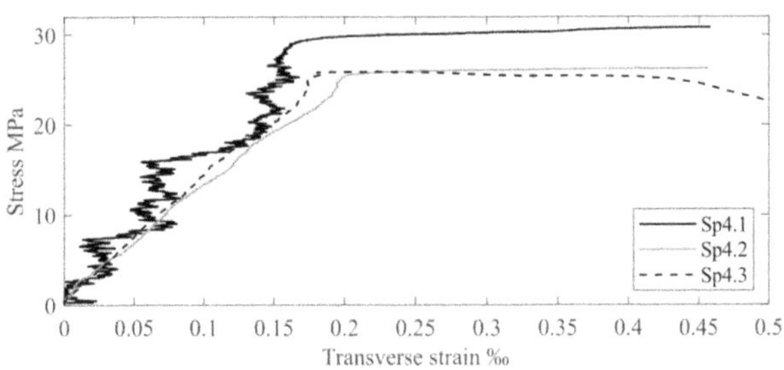

Fig. 4 Experimental dependencies σ_c vs. ε_{trans}, obtained in the second test series.

6.2 Cylindrical Specimen Behavior in Inelastic State

Following modern design codes [1, 2], Poisson's ratio determines concrete behavior in transverse direction. At the same time, based on the experimental results obtained in the framework of the present study, it is very important to consider the transverse deformations over the range of Poisson's ratio that is usually used. With this aim, the dependence of transverse deformations on longitudinal ones was analyzed. The relationship between these deformations for the first test series is shown in Figure 5. As it is evident from this figure, the dependence is non-linear in the interval when $\varepsilon_{trans} > \varepsilon_{ct\,ul}$ (see the values of $\varepsilon_{ct\,ul}$ in Table 1).

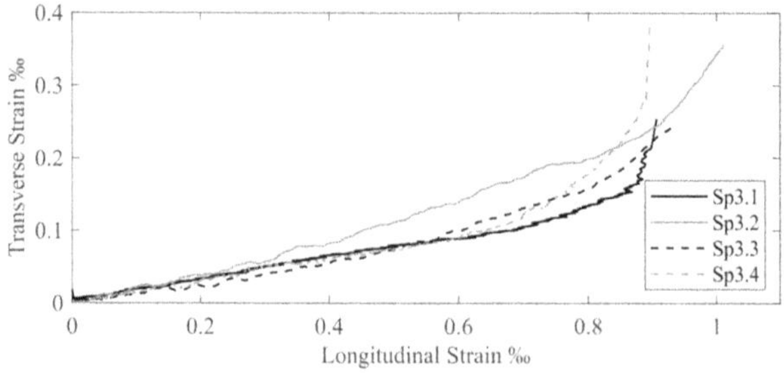

Fig. 5 Dependencies of transverse strains on longitudinal ones.

Table 3 presents the values of Poisson's ratios as well as the relationship between the maximum transverse deformations, $\varepsilon_{trans\,1}$, and longitudinal ones, ε_{c1}, at ULS. Following this table, the average ratio between these two parameters and the Poisson's ratio, $\frac{\frac{\varepsilon_{trans\,1}}{\varepsilon_{c1}}}{\nu}$ is about 1.8. It should be noted that, in accordance with the Structural Phenomenon [60], this value should be close to 2 [73]. The difference between the obtained experimental and theoretical results is about 10%, which is explained by concrete heterogeneity and the probable technological eccentricity that was described in Section 4.5.3.

Table 3 Relationship between transverse and longitudinal strains at ULS and Poisson's ratio values.

Specimen	ν	$\frac{\varepsilon_{trans\,1}}{\varepsilon_{c\,1}}$	$\frac{\frac{\varepsilon_{trans\,1}}{\varepsilon_{c\,1}}}{\nu}$
Sp3.1	0.15	0.28	1.86
Sp3.2	0.22	0.35	1.57
Sp3.3	0.16	0.26	1.63
Sp3.4	0.15	0.32	2.13
Average	0.17	0.30	1.80

The dependence between $\frac{\varepsilon_{trans}}{\varepsilon_c}$ and the longitudinal deformations, ε_c, is shown in Figure 6. This dependence is an additional tool that demonstrates the correspondence of concrete behavior to the previously developed theoretical model [74]. Following Figure 6, the value of $\frac{\varepsilon_{trans}}{\varepsilon_c}$ increases very slowly approximately up to $\varepsilon_c = 0.9‰$, and then a sharp increase in this ratio is evident. It also confirms that concrete behavior in this area is inelastic and corresponds to the theoretical model. According to this model, the ultimate elastic concrete behavior is limited by a longitudinal strain value of 1.0‰, which correlates with the experimental value of 0.9‰. At the longitudinal crack development stage, caused by transverse deformations, the ratio $\frac{\varepsilon_{trans}}{\varepsilon_c}$ tends to be 0.4, which is approximately double as compared to the elastic stage. This statement fully corresponds to the Structural Phenomenon [60].

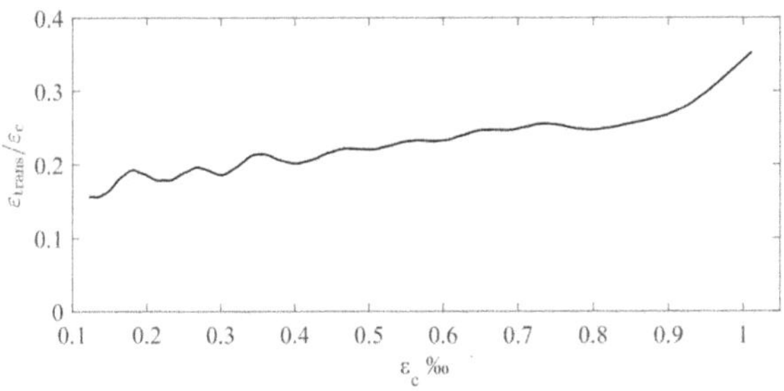

Fig. 6 Dependencies between average $\frac{\varepsilon_{trans}}{\varepsilon_c}$ and ε_c

6.3 Analysis of Specimen Failure Progress Caused by Transverse Deformations

To analyze the entire process of longitudinal crack development, frame-by-frame images were used (Figure 7). This process is divided into three stages [73]. At the first stage, when $\varepsilon_{trans} \leq \varepsilon_{ct\ ul}$, compressed concrete is exposed to transverse deformations until initiation of longitudinal cracks as a result of compression in the longitudinal direction (Figure 7a). According to the results presented in Table 1, the average value of $\varepsilon_{ct\ ul}$ for the three specimens is equal to 0.095‰.

At the second stage, when $\varepsilon_{ct\ ul} < \varepsilon_{trans} \leq 2\varepsilon_{ct\ ul}$, i.e., transverse deformations develop until about 0.2‰ (Figure 2), there is a process of longitudinal crack initiation and development (Figure 7b). At the selected loading rate, this stage continues for about 6 seconds. At the third stage, the crack propagation process ends with specimen failure, when

1.0 s
(a)

2.0 s 3.0 s 4.5 s
(b)

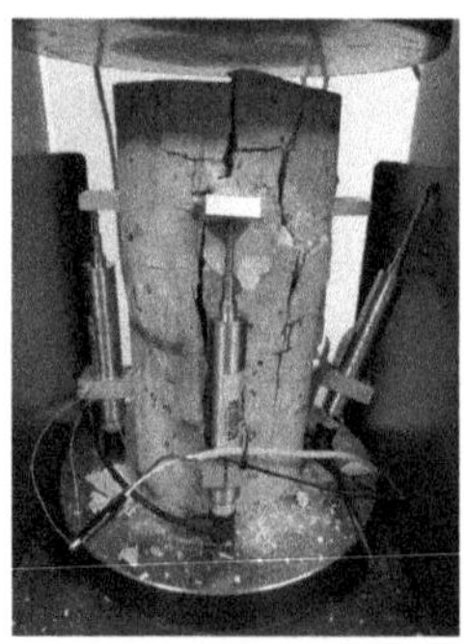

After 6 s
(c)

Fig. 7 Typical crack propagation in the investigated specimens: (a) End of stage 1 – crack development initiation; (b) Stage 2 – crack development process; (c) Stage 3 – specimen failure.

$\varepsilon_{trans} > 2\varepsilon_{ct\ ul}$ (Figure 7c). The obtained longitudinal crack development pattern corresponds to the deformation process and failure in the columns. For example, a similar behavior in the reinforced concrete columns of residential buildings was observed during the Mexico earthquake in 2017 [73].

Crack propagation in the specimens occurs parallel to the compressive force direction that proves deformation development in the transverse direction, which is caused only by pure tension due to the cylindrical form of the specimens. As a result, the longitudinal crack development process in specimens leads to their separation into several vertical wedge-shaped segments and an unbroken cylindrical inner core, which was predicted theoretically earlier [73].

The following important investigation step is to analyze the specimen failure type after unloading (Figure 8). Following the Structural Phenomenon [60], it can be theoretically predicted that the cross-section area of the destroyed specimens is approximately half that of the initial one. Experimental studies confirmed this prediction for all of the tested

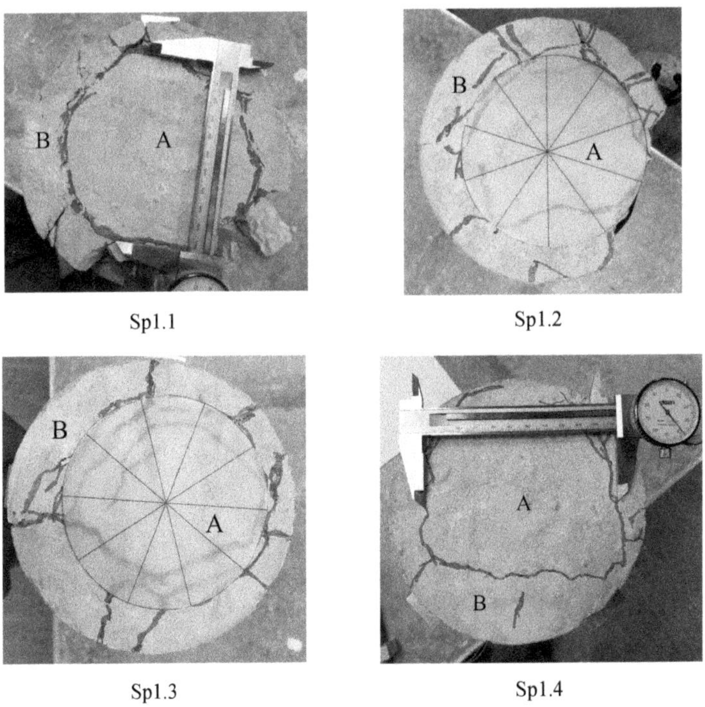

Fig. 8 Failure pattern of the upper surface of the specimens.

specimens. The area of the inner core cross-section A is about twice lower compared to the outer segments B (Figure 8), where the sum of A and B is equal to the initial cross-section of the specimen. The ratios between the entire cross-section area of each specimen and that of the inner core A are presented in Table 4. The obtained data show that the average ratio is close to 2.

Table 4 Measured values of $(A + B)$ and A and their ratio.

Specimen	*(A + B), cm²*	*A, cm²*	$\frac{(A+B)}{A}$
Sp1.1	176.5	91.6	1.9
Sp1.2	176.4	75.1	2.3
Sp1.3	176.4	74.9	2.4
Sp1.4	177.1	99.4	1.8
Average	176.6	85.3	2.1

The typical failure pattern of a compressed concrete specimen is presented in Figure 9.

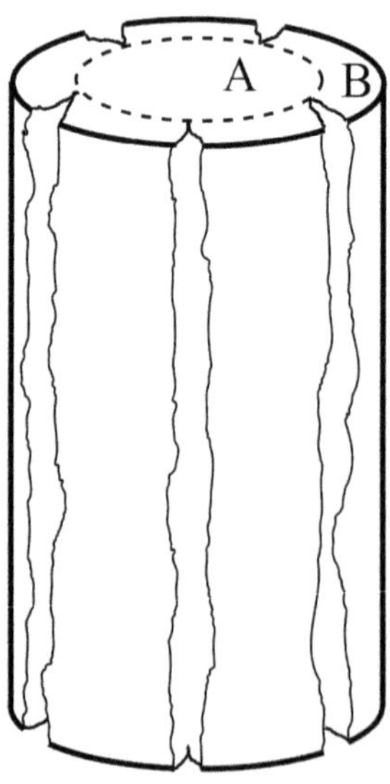

Fig. 9 Typical failure pattern of compressed concrete specimen.

Reliability of results obtained for all seven specimens within the framework of the present research phase is also confirmed by available in the literature data: eight cylindrical NSC specimens with a height-to-diameter ratio equal to 2 were tested under uniaxial loading [69]. The diameters of the specimens differed from 10 to 80 cm. It was reported that crack development in the vertical direction and formation of an unbroken cylindrical inner core corresponds to the described failure pattern (see Figure 2: 20-40-02 and 40-80-02 in [69]).

Thirty-three 15 × 30 cm fibered-concrete cylindrical specimens with a minimum 28-day concrete compressive strength of 27.6 MPa were axially loaded [70]. The typical failure pattern with vertical longitudinal cracks is presented in Figure 8 in [70]. The obtained stress-longitudinal strain relationships for compressed concrete specimens are linear, almost until peak load (Figure 15 in [70]). Thus, it should be mentioned that the results obtained for concrete mixes with different amounts and types of fiber correspond to the results obtained in the present study for NSC specimens.

Six cylindrical NSC specimens with different end conditions were subjected to uniaxial compression [71]. In this case, transverse deformations were additionally measured. The obtained data show a good convergence with the present research results. The stress-longitudinal strain relationships are linear until failure when the stress-transverse strain relationships have well pronounced non-elastic behavior until a peak load (see Figures 2 and 3 in [71]). The earlier described failure pattern is obtained for specimens with reduced friction conditions (see Figure 5 in [71]).

6.4 Comparison of Normal and High Strength Compressed Concrete Behavior

Based on the data, obtained in the framework of the present study, the behavior of NSC, which corresponds to class C30 at relatively high loading rates, can be compared with that of HSC [73]. The available data on HSC (class C90) includes test results for three cylindrical specimens. Longitudinal deformations were measured by strain gauges using linear variable displacement transducers (LVDT) and transverse deformations. The dependencies of these two deformations' types vs. stresses under axial load are shown in Figures 10 and 11, respectively. Corresponding data for comparison of elastic and non-elastic behavior of HSC and NSC are provided in Tables 5 and 6.

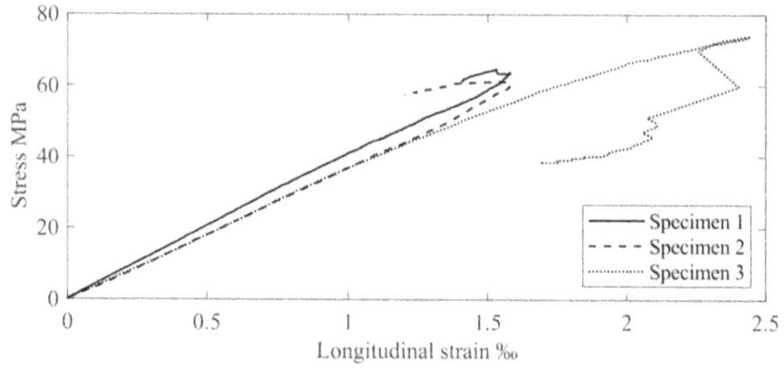

Fig. 10 Experimental $\sigma_c - \varepsilon_c$ dependencies for HSC (following [73]).

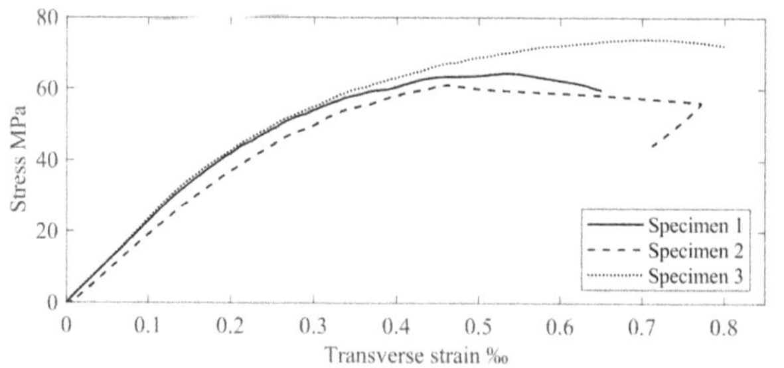

Fig. 11 Experimental $\sigma_c - \varepsilon_{trans}$ dependencies for HSC (following [73]).

Table 5 Experimental values of maximum elastic strains, ultimate transverse strains, and Poisson's ratios for HSC (following [73]).

Specimen	$\varepsilon_{c\,el}$, ‰	$\varepsilon_{ct\,ul}$, ‰	v
Specimen 1	0.780	0.143	0.18
Specimen 2	0.829	0.161	0.19
Specimen 3	1.004	0.162	0.16
Average	0.871	0.155	0.18

Table 6 Relationships between experimental Poisson's ratio values, transverse, and longitudinal strains at ULS (following [73]).

Specimen	v	$\frac{\varepsilon_{trans1}}{\varepsilon_{c1}}$	$\frac{\varepsilon_{trans1}/\varepsilon_{c1}}{v}$
Specimen 1	0.18	0.34	1.89
Specimen 2	0.19	0.29	1.53
Specimen 3	0.16	0.29	1.81
Average	0.18	0.31	1.74

The transverse deformation limit of HSC specimens at concrete elastic behavior is 0.15‰. In this case, the external load value is approximately half of the concrete strength, which corresponds to the Structural Phenomenon [60]. The average Poisson's ratio is equal to 0.18. Further load increases are accompanied by the linear development of longitudinal deformations (Figure 10), while the transverse deformations begin to exhibit elastic-plastic behavior (Figure 11). Therefore, it is logical to analyze the ratio between the longitudinal and transverse deformations for further interpretation of HSC behavior. Following Table 6, at ULS, this ratio is equal to 0.31, which is close to the results obtained in the present study (Table 3). It should be mentioned that the growth of relative transverse deformations, $\frac{\frac{\varepsilon_{trans1}}{\varepsilon_{c1}}}{v}$, in both cases (for HSC and NSC) is equal to about 1.7.

Comparison of results obtained for NSC and steel-fibered HSC (Tables 3 and 6) shows sufficient convergence of the standard deviations for Poisson coefficient. The standard deviations for $\frac{\varepsilon_{trans1}}{\varepsilon_{c1}}$ are 0.024 and 0.031, respectively, and the variation coefficients are 13.7% and 10.1%, correspondingly. Thus, the obtained data confirm that it is possible to use experimental results for NSC specimens obtained at rather a high loading rate for interpretation of HSC behavior. It allows disclosure of the non-elastic behavior stage in compressed steel fibered HSC, which is important for developing proper and effective design approaches.

7

Numerical Investigation of Transverse Cracks' Depth and Width Development in Compressed Concrete Elements

7.1 Hypotheses

Crack width and depth development due to transverse deformations in compressed concrete elements under axial loading was previously described and the final state of the tested specimen is presented in Figure 9 (Chapter 6). At the present stage, numerical step-by-step analysis is proposed for further deeper investigation of this process. The following basic hypotheses are used:

- the crack shape in horizontal section has an isosceles triangle shape;
- Each calculation step of the crack width, Δw, and depth, Δh, is determined as a function of longitudinal deformation development;
- the triangle angles, α and β, remain constant from step-to-step (see Figures 1 and 2).

7.2 Overview of Experimental Data Step Analysis

At each step, i, the analysis is based on calculation of crack width and depth, w_i and h_i ($i = 1...5$), defining undamaged circular element width in cylindrical specimen section, t_i, in each calculation step (Figures 1 and 2). Figure 1 presents the calculation algorithm diagram. Schemes of the performed analysis are demonstrated in Figure 2.

As shown in the algorithm diagram (Figure 1), the calculation runs in the **reverse** order, starting from the specimen limit state ($\sigma_c = f_c$ and $\varepsilon_c = \varepsilon_{c1}$), when the section area of the specimen inner undamaged core

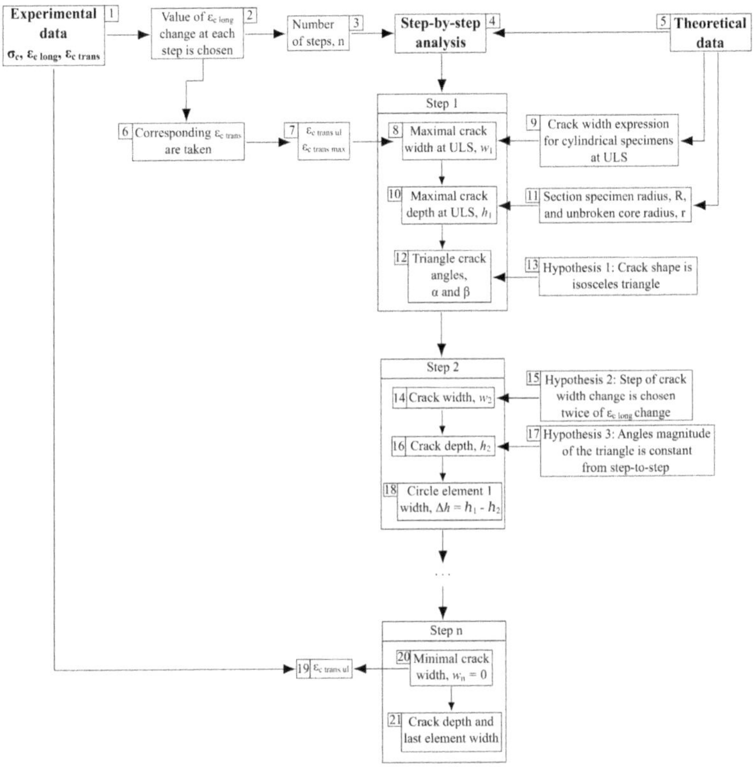

Fig. 1 Crack width and depth calculation algorithm in compressed concrete element.

is equal to half of its initial value [74] (see cell 11 in Figure 1) (see also Figure 2a). The calculation ends at the stage corresponding to the crack initiation due to transverse deformations (the end of specimen elastic behavior), when the following parameters are obtained:

- crack width is theoretically equal to zero;
- experimental value of transverse deformation corresponds to concrete tensile strength in the transverse direction, f_{ct} [112] (cells 19–21 in Figure 1).

At the initial calculation stage, preceding to the calculation of Step 1, a certain longitudinal compression strains' step is selected equal to $\frac{\varepsilon_{c\,long\,i}}{1.44}$ (1.44 is selected to obtain an integer number of calculation steps), based on their limit experimental value, ε_{c1} (see cell 2 in Figure 1).

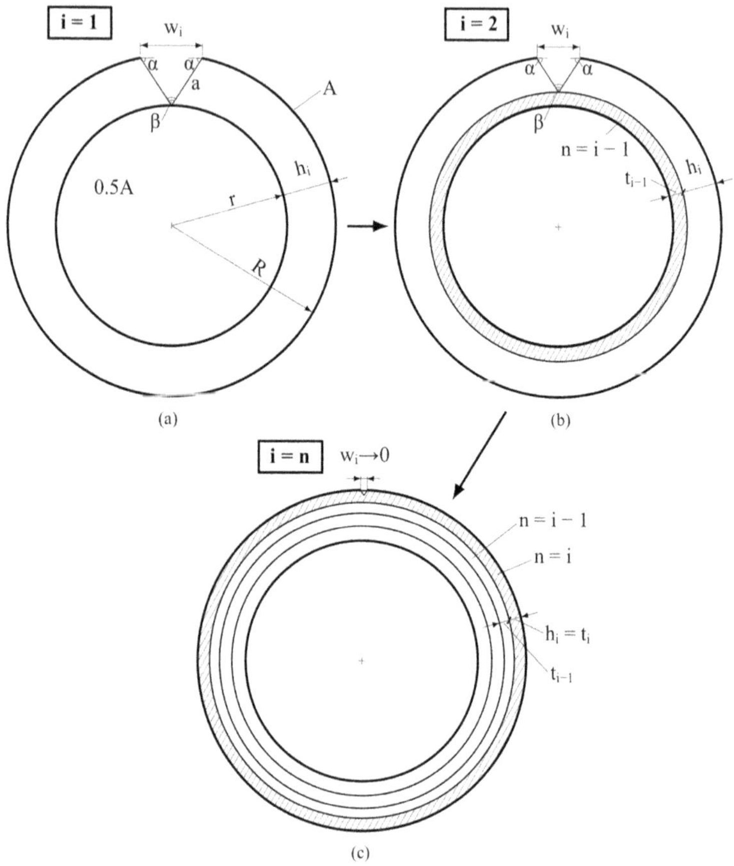

Fig. 2 Step-by-step numerical analysis schemes: (a) Step 1 – cracked specimen at the peak stress (**limit** loading stage); (b) Step 2 – the first circle element calculation (**intermediate** loading stage); (c) Final step – the last circle element calculation when the crack width tends to zero (loading stage corresponding to the **end of elastic behavior**).

It determines the calculation steps number n obtained using geometric progression (see cell 3 in Figure 1):

$$n = 1 + \frac{\log \dfrac{\varepsilon_{c\,el}}{\varepsilon_{c1}}}{\log \dfrac{1}{1.44}} \qquad ...(1)$$

For example, for limit and maximal elastic longitudinal deformation values, $\varepsilon_{c1} = 2‰$ and $\varepsilon_{c\,el} = 0.5‰$ respectively, the number of steps $n = 5$.

Further, according to longitudinal strain experimental values $\varepsilon_{c\ long}$ the corresponding transverse strains $\varepsilon_{c\ trans}$ are taken (see cell 6 in Figure 1). Thus, the basic calculation parameters are determined (see cells 1–3 and 6–7 in Figure 1).

At the next stage, the first step of analysis, (i = 1), starts (see cells 8, 10, and 12 in Figures 1 and 2a). At this step, the specimen limit state by transverse deformations is considered. At the beginning, maximum specimen crack width $w_1 = w_{\max}$ is calculated using known expression, describing concrete tensile behavior at different stress-strain conditions (see Eq. (5) in [112]). The crack width calculation is based on ultimate transverse strains $\varepsilon_{c\ trans\ ul}$ and the transverse strain corresponding to maximal crack width at peak stress $\varepsilon_{c\ trans\ \max}$ (see cells 7–9 in Figure 1). Next, the crack depth $h_1 = h_{\max}$ is calculated: it is equal to the difference between the specimen section radius before loading, R, and after reaching the limit state in longitudinal direction (which is equal to undamaged core radius), r (see cells 10 and 11 in Figures 1 and 2a). After that, triangle angles α and β are calculated based on the first hypothesis, which determines the crack shape in the horizontal section as an isosceles triangle (see cells 12 and 13 in Figures 1 and 2a).

The second step of the analysis (i = 2) starts from calculation of the crack width $w_2 < w_1$ (because of the calculation reverse order), using the second hypothesis (see Figure 2b and cells 14 and 15 in Figure 1). According to this hypothesis, the incremental change in crack width is assumed to be twice the longitudinal strains change (according to the Structural Phenomenon [60]). In a general case, the expression for crack width calculation is:

$$w_i = \frac{w_{i-1}}{(2 \cdot 1.44)} \qquad ...(2)$$

Next, the crack depth, $h_2 < h_1$, is calculated based on the third hypothesis (see cells 16 and 17 in Figures 1 and 2b). After that, the width of element 1, t_1, is calculated – it is equal to the difference between the crack depths from steps 1 and 2 (see cell 18 in Figures 1 and 2b). In general, the equation for element width calculation has the following form:

$$t_{i-1} = h_{i-1} - h_i \qquad ...(3)$$

The process described in Step 2 is repeated (n – 1) times. The step-by-step analysis ends by the step n (Figure 2c), at which the calculated crack width is theoretically equal to zero and the transverse strain value is equal to $\varepsilon_{c\ trans\ ul}$ (see cells 19–21 in Figure 1).

7.3 Numerical Analysis of the Cracks' Depth and Width Development Based on the Experimental Transverse Deformations

As described previously, the available experimental data on longitudinal and transverse deformations for compressed concrete cylindrical specimen [113] are used in the present step-by-step analysis. Relationships between stresses and longitudinal strains as well as stresses and transverse strains (including basic points 1, 3, 4, 5, described previously in Figure 20 [Chapter 2]) are shown in Figures 3 and 4.

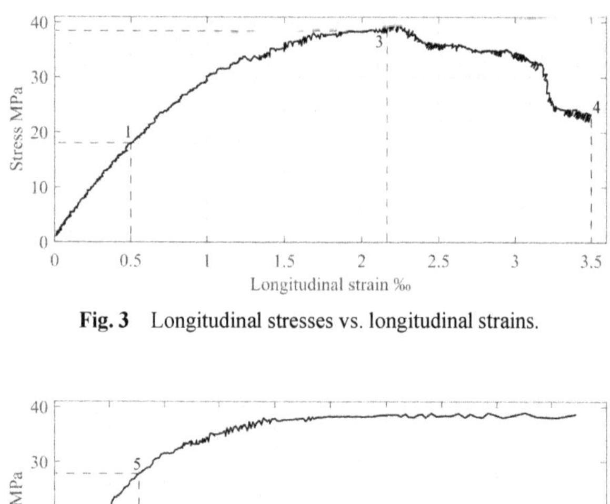

Fig. 3 Longitudinal stresses vs. longitudinal strains.

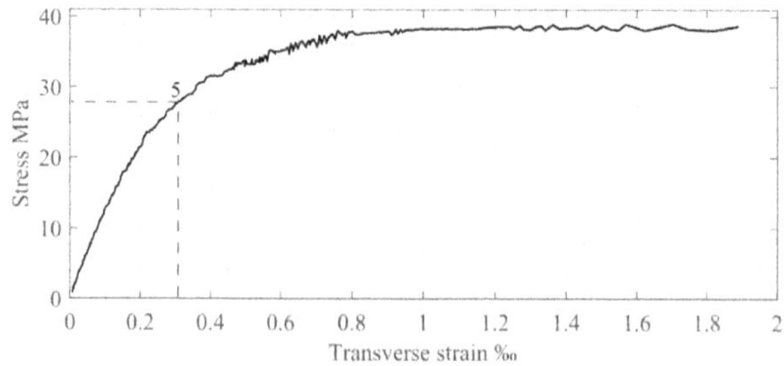

Fig. 4 Longitudinal stresses vs. transverse strains.

To obtain a more accurate representation of compressed concrete behavior, it is proposed to combine the relationships as shown in Figures 3 and 4 (Figure 5). The obtained diagram demonstrates the relation of the following three dependencies: longitudinal and transverse strains development in concrete cylindrical specimen (σ_c vs. ε_c, σ_c vs. $\varepsilon_{c\,trans}$, and ε_c vs. $\varepsilon_{c\,trans}$), depending on compression loading.

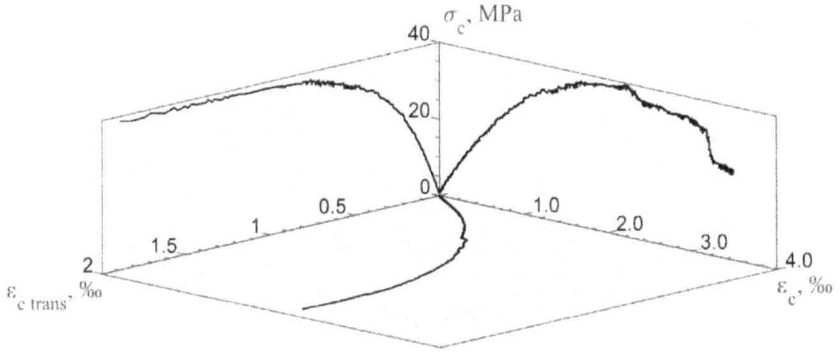

Fig. 5 Experimental relationship between longitudinal stresses and longitudinal and transverse strains.

Table 1 presents selected experimental data for the initial step-by-step analysis stage. As mentioned above, the number of calculation steps is selected to be equal to 5, when Step 1 corresponds to the concrete limit state. Additionally, the modulus of deformability $E_{c\ deform}$ is calculated for each step, based on the experimental data presented in Table 1. The result obtained at Step 1 demonstrates a twofold decrease in the modulus magnitude by reaching the peak stress (from 36.1 to 18 MPa), which corresponds to the Structural Phenomenon.

Table 1 Basic experimental data.

Step Number	σ_c, *MPa*	ε_c, ‰	$E_{c\ deform}$, *MPa*	$\varepsilon_{c\ trans}$, ‰
1	38.96	2.16	18.0	1.703
2	23.88	0.72	33.1	0.230
3	30.63	1.04	29.5	0.373
4	23.88	0.72	33.1	0.230
5	18.05	0.50	36.1	0.156

The experimental data presented in Table 1 is used for the step-by-step numerical analysis. According to the above described analysis algorithm (Figure 1), the following parameters of compressed concrete behavior are calculated at each step: crack width w_i, crack depth h_i, and element width t_i. The results are presented in Table 2, which includes comparison of crack widths, w^{Theor} and w^{Exper}, calculated using theoretical deformation values [112] and experimental ones (Table 1), respectively. Graphical representation of stress and crack width relationship is shown in Figure 6.

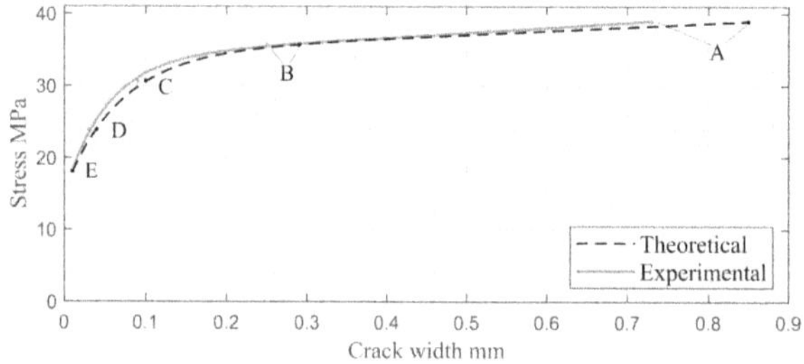

Fig. 6 Stresses vs. crack widths due to transverse deformations: A – step 1; B – step 2; C – step 3; D – step 4; E – step 5.

At Step 1 the crack width corresponds to the element limit state. In this case the calculation of w^{Theor} is based on transverse deformation $\varepsilon_{c\,trans\,\max}$: 2 – 0.2 = 1.8‰, where 0.2‰ is the theoretical value of ultimate transverse deformations $\varepsilon_{c\,trans\,ul}$ and 2‰ is the transverse deformation value corresponding to maximal crack width. These values correspond to the experimental transverse deformations at steps 5 ($\varepsilon_{c\,trans}$ = 0.156‰) and 1 ($\varepsilon_{c\,trans}$ = 1.703‰) (Table 1), which were used for determining w^{Exper}. The obtained relative error D = 14.02% shows rather good convergence between theoretically and experimentally based results for crack width at limit state of compressed concrete element. The same numerical data is shown in Figure 7.

The maximum crack depth $h_{\max} = h_1$ is also defined at Step 1 as equal to 22 mm (Table 2 and Figure 7). This value is calculated according to the above described experimentally verified theoretical model of concrete cylindrical specimen section at ULS, based on the Structural Phenomenon [60] (specimen section area is double compared to that of inner undamaged core). Correspondingly, the sum of all element widths is:

$$\sum_{i=1}^{n} t_i = h_1 \qquad ...(4)$$

Table 2 Numerical results of the step-by-step analysis.

Step number	w^{Theor}, *mm*	w^{Exper}, *mm*	*D, %*	*h, mm*	*t, mm*
1	0.85	0.73		22.00	14.36
2	0.30	0.25		7.64	4.99
3	0.10	0.09	14.02	2.65	1.73
4	0.04	0.03		0.92	0.60
5	0.01	0.01		0.32	0.32

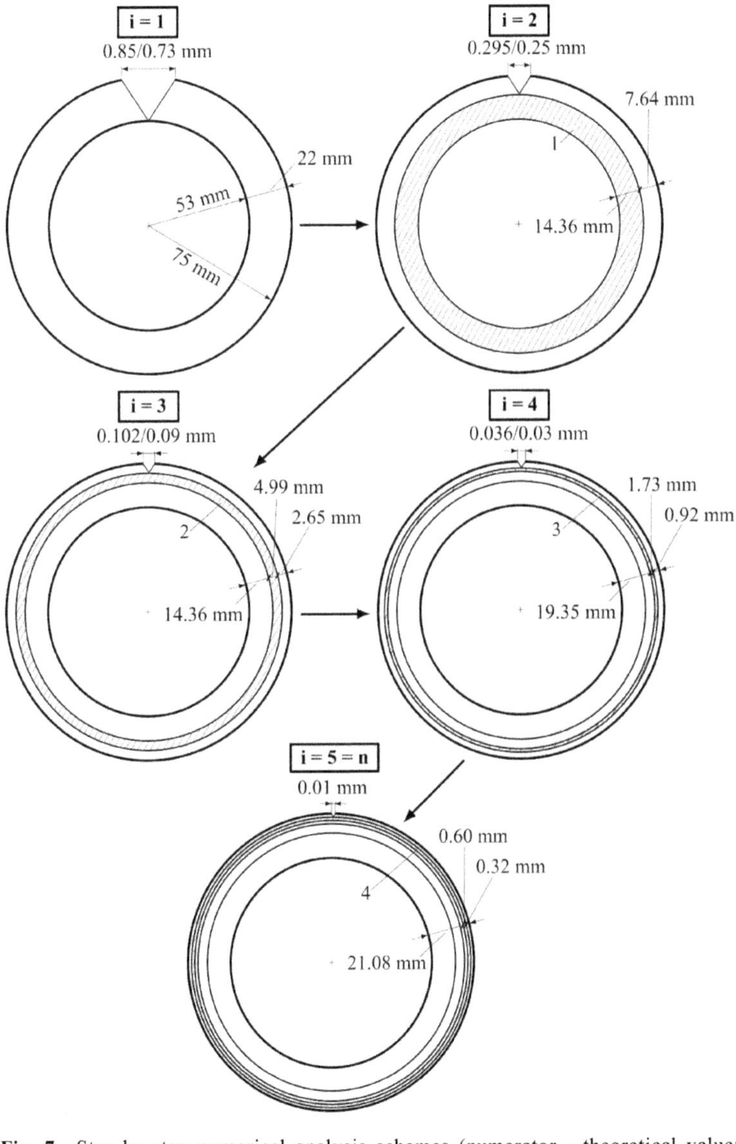

Fig. 7 Step-by-step numerical analysis schemes (numerator - theoretical value; denominator - experimental one).

According to the obtained numerical results, the crack width reaches 0.3 mm at Step 2 (this step is the pre-ultimate stage before the element limit state). This value complies with the modern code requirements for the maximum crack width in concrete elements [1, 2]. At the last calculation step ($\sigma_{ct\ trans} = f_{ct}$) the crack width tends toward zero when transverse deformations correspond to crack initiation (Table 1, Figures 5 and 6).

Thus, the complex numerical analysis of compressed concrete element behavior was conducted, based on investigation of crack development, caused by transverse deformations. The research outcomes justify the proposed approach efficiency for investigation of compressed concrete at limit state. With this aim are used appropriate parameters of concrete behavior, not only in the longitudinal direction but also in the transverse one.

8

Compressed Concrete Elements Limit States by Width and Depth of Cracks Due to Transverse Deformations

The present section provides an example that is aimed at identifying the limit state of a cylindrical compressed concrete specimen with 300 mm height and 150 mm diameter by transverse deformations. The following theoretical data, as well as the experimentally obtained one, is used for step-by-step analysis:

- $\varepsilon_{c\ trans\ ul}^{Theor} = 0.200‰$,
- $\varepsilon_{c\ trans\ ul}^{Exper} = 0.156‰$,
- $\varepsilon_{c\ trans\ \max}^{Theor} = 2.000‰$,
- $\varepsilon_{c\ trans\ \max}^{Exper} = 1.700‰$,
- $R = 75$ mm,
- $r = 53$ mm.

At the first analysis step ($i = 1$), the maximum specimen crack width is calculated:

$$w_1^{Theor} = \frac{(2.000 - 0.200) \cdot \pi \cdot 150}{1000} = 0.85 \text{ mm} \qquad ...(1)$$

$$w_1^{Exper} = \frac{(1.700 - 0.156) \cdot \pi \cdot 150}{1000} = 0.73 \text{ mm} \qquad ...(2)$$

The maximum crack depth is:

$$h_1 = 75 - 53 = 22 \text{ mm} \qquad ...(3)$$

The triangle angles of the crack shape α and β are calculated, when the hypotenuse lengths (Figures 1 and 2 [Chapter 7]) are

$$a^{Theor} = 22.0041 \text{ mm and } a^{Exper} = 22.0030 \text{ mm}:$$

$$\alpha = \left(\frac{w}{2a}\right); \quad \beta = 180° - 2\alpha \qquad ...(4)$$

The obtained results are shown in Table 1.

Table 1 Theoretical and experimental values of triangle angles, α and β.

	Theoretical, °	*Experimental,* °
α	88.90	89.05
β	2.21	1.90

At the second analysis step (i = 2), the crack width is calculated according to Eq. (2 [Chapter 7]):

$$w_2^{Theor} = \frac{0.85}{2 \cdot 1.44} = 0.30 \text{ mm} \qquad ...(5)$$

$$w_2^{Exper} = \frac{0.73}{2 \cdot 1.44} = 0.25 \text{ mm} \qquad ...(6)$$

The crack depth at this step is:

$$h_2 = \frac{w_2}{2\cos\alpha} \cdot \cos\left(\frac{\beta}{2}\right) = 7.64 \text{ mm} \qquad ...(7)$$

The first element width is calculated using Eq. (3 [Chapter 7]):

$$t_1 = 22 - 7.64 = 14.36 \text{ mm} \qquad ...(8)$$

Steps 3, 4, and 5 of the analysis repeat procedures are described for Step 2. The results of the analysis are presented in Table 2 (Chapter 7).

Thus, in the calculation example the final theoretical and experimental values of crack width and depth in the compressed concrete element are obtained at all stages of its performance, including the limit state. The next stage is to compare the specified values with the modern code requirements for final formulation of the limit state calculation by transverse deformations in a real element.

It should be noted that the compressed element limit state by the crack depth, caused by transverse deformations, is proposed for the first time and should be associated with a protective coating of reinforcement. The obtained maximum crack depth in the example is equal to 7.64 mm, which corresponds to the maximum crack width value, required by modern codes [1, 2]. This depth value is usually acceptable for compressed elements, for which concrete coating layer is at least 1 cm. This parameter is especially critical for thin-walled RC structures, in particular for shells.

Conclusions

Complex analysis of compressed concrete limit states was conducted based on the existing theoretical and experimental data, including longitudinal and transverse deformations, as well as longitudinal cracks' width and depth. Based on the obtained results the following conclusions are evident:

1. The following main ideas of the theoretical compressed concrete stress-strain relationship were experimentally confirmed:
   - stresses and strains correspond to the border between the elastic and inelastic compressed concrete behavior;
   - compressed concrete elastic potential (important for dynamic analysis and investigations of reinforced concrete structure creep);
   - geometrical correspondence between the descending the ascending branches;
   - based on the Structural Phenomenon, the descending branch limit stress is equal to $0.5f_c$.
2. It was for the first time demonstrated that the experimental investigations must completely confirm the stress-strain diagram of compressed concrete.
3. Consequently, this diagram was called the "Law of Concrete" and should be included in the modern concrete theory. The law ensures proper and accurate design of compressed reinforced concrete elements without empirical coefficients, which is significant for further development of structural concrete theory.
4. The conducted experiments revealed the importance of considering the testing machine deformations in obtaining the stress-strain diagram descending branch. It was associated with the stiffness of the testing machine, which affects the specimen deformations.
5. The importance of the experimental investigation of normal strength concrete at relatively high loading rates (in order to prove the effectiveness of using transverse deformations as an indicator for

the interpretation of high-strength concrete behavior) was verified, which simplifies experimental investigation of high strength concrete. It was confirmed that from the viewpoint of transverse deformations, high-strength concrete also behaves in elastically as a normal strength one. In the author's opinion, further investigation of this phenomenon can form a basis for more accurate and effective design of high-strength concrete structures.

6. The theoretical failure pattern for compressed cylindrical concrete specimens, which includes an unbroken circular core, surrounded by separate vertical wedge-shaped elements, was experimentally confirmed. The ratio between the initial specimen's cross-section area and the unbroken one was experimentally obtained as approximately equal to 2, which corresponds to the Structural Phenomenon.
7. A new approach for simultaneous consideration of compressed concrete element limit states by transverse deformations was proposed. A corresponding graph of axially loaded concrete behavior in longitudinal and transverse directions was used for this analysis.
8. A reverse order step-by-step analysis was proposed, starting from compressed concrete element ULS and finishing at the end of its elastic stage. At the initial stage it allowed obtaining the crack width and depth caused by transverse deformations.
9. The limit state by crack depth was introduced as a function of crack width development. It allows theoretical determination of the minimum concrete coating value for reinforcement in compressed elements, which is currently proposed in modern codes as an empirical value.

References

[1] British Standards Institution, *EN 1992-1-1: Eurocode 2: Design of Concrete Structures Part 1-1: General Rules and Rules for Buildings,* London, UK: British Standards Institution, 2004.

[2] The Standards Institution of Israel, *SI 466 Part 1: Concrete Code: General Principles,* Tel Aviv, Israel: The Standards Institution of Israel, 2017.

[3] I. Iskhakov and Y. Ribakov, "Theoretical Stress-Strain Model for Compressed Composite Cement Materials," in *Transaction High Performance and Optimum Design of Structures and Materials III*, Southampton, UK, 2018.

[4] E. Hognestad, *Study of Combined Bending and Axial Load in Reinforced Concrete*, vol. 399, Champaign, IL: University of Illinois at Urbana Champaign, 1951, p. 128.

[5] M. Koenen, "Für die Berechnung der Stärke der Monierschen Cementplatten," *Zentralblatt der Bauverwaltung,* vol. 6, no. 47, p. 462, 1886.

[6] P. Neumann, "Ueber die Berechnung der Monier-Constructionen," *Wochen schrift des Österr. Ing. Und Arch. Vereins,* vol. 15, no. 22, pp. 209–212, 1890.

[7] E. Coignet and N. d. Tedesco , "Du calcul des Ouvrages en Ciment avec Ossature Métallique," *Mémoires de la Société des Ingénieurs Civils de France,* vol. I, pp. 282–363, 1894.

[8] J.B. Johnson, "Strength of Concrete and Steel in Combination," *Engineering News and American Railway Journal,* vol. 33, no. 1, pp. 10–11, 1895.

[9] M.R. v. Thullie, "Ueber die Berechnung der Biegespannungen in den Beton- und Monier-Constructionen," *Zeitschrift des Oesterr. Ing. Und Arch. Vereins,* vol. 48, no. 24, pp. 365–369, 1896.

[10] M.R. v. Thullie, "Ueber die Berechnung der Monierplatten," *Zeitschrift des Oesterr. Ing. Und Arch. Vereins,* vol. 49, no. 13, pp. 193–197, 1897.

[11] F.v. Emperger, "Zur Theorie der Verstärkten Betonplatte," *Zeitschrift des Oesterr. Ing. Und Arch. Vereins,* vol. 49, no. 22, pp. 351–355 and 364–367, 1897.

[12] W. Ritter, "Die Bauweise Hennebique," *Schweizerische Bauzeitung,* vol. 33, nos. 5, 6, and 7, pp. 41–43, 49–52, and 59–61, 1899.

[13] A. Ostenfeld, "Einige Folgerungen aus den Sanders'schen Belastungsversuchen," *Beton und Eisen,* vol. 1, no. 5, pp. 19–22, 1902.

[14] A.N. Talbot, *Tests of Reinforced Concrete Beams*, vol. 1, Champaign, IL: University of Illinois, Engineering Experiment Station Bulletin, 1904, p. 64.

[15] A.N. Talbot, *Tests of Reinforced Concrete Beams Series of 1905*, vol. 4, Champaign, IL: University of Illinois, Engineering Experiment Station Bulletin, 1906, p. 84.

[16] C. Bach and O. Graf, "Tests with Simply Supported, Quadratic Reinforced Concrete Plates," *Deutscher Ausshuss für Eisenbeton,* vol. 30, 1915.

[17] W.M. Wilson, F.E. Richart, and C. Weiss, *Analysis of Statically Indeterminate Structures by the Slope-Deflection Method*, vol. 108, Champaign, IL: University of Illinois, Engineering Experiment Station Bulletin, 1918, p. 214.

[18] H. Cross, *The Column Analogy*, vol. 215, Champaign, IL: University of Illinois, Engineering Experiment Station Bulletin, 1930, p. 75.

[19] H. Cross and N. D. Morgan, *Continuous Frames of Reinforced Concrete*, Hoboken, NJ: John Wiley and Sons, Inc., 1932, p. 343.

[20] M. Roš, "Versuche und Erfahrungen an ausgeführten Eisenbeton-Bauwerken in der Schweiz," *Eidgenössische Materialprüfungs- und Versuchsanstalt für Industrie*, Bauwesen und Gewerbe, Zürich, 1937.

[21] W.M. Wilson, R.W. Kluge, and J.V. Coombe, *An Investigation of Rigid Frame Bridges*, Part II, vol. 308, Champaign , IL: University of Illinois, Engineering Experiment Station Bulletin, 1938, p. 75.

[22] F.R. McMillan, "A Study of Column Test Data," *Proceedings ACI*, vol. 17, pp. 150–171, 1921.

[23] ACI Journal, *ACI 318-47: Building Code Requirements for Reinforced Concrete*, vol. 44, ACI Journal, 1947, pp. 1–64.

[24] ACI Committee 318, *ACI CODE-318-19(22): Building Code Requirements for Structural Concrete and Commentary*, Farmington Hills, MI: American Concrete Institute, 2019.

[25] E. Suenson, "Jærnprocentens Indflydelse paa Jænbetonpladers Bæreevne," *Ingenøiren, Copenhagen*, vol. 21, no. 72, p. 568, 1912.

[26] G.v. Kazinczy, "Die Plastizität des Eisenbetons," *Beton und Eisen*, vol. 32, no. 5, pp. 74–80, 1933.

[27] L.J. Mensch, "New-Old Theory of Reinforced Concrete Beams in Bending," *ACI Journal*, vol. 2, no. 7, pp. 28–41, 1914.

[28] W.E. Ritter, "A Step Forward in the Methodology of Natural Science," in *Proceedings of the American Academy of Arts and Sciences*, vol. 55, no. 2, pp. 91–133, 1919.

[29] H. Kempton Dyson, "What is the Use of the Modular Ratio?" *Concrete and Constructional Engineering*, vol. 17, nos. 5, 6, and 7, pp. 330–336, 408–415, 486–491, 1922.

[30] F. Stüssi, "Ueber die Sicherheit des einfach bewehrten Eisenbeton-Rechteckbalkens," *Publications, International Association for Bridge and Structural Engineering*, vol. 1, pp. 487–495, 1932.

[31] S. Steuermann, "Das Wiederstandsmoment eines Eisenbetonquerschnittes," *Beton und Eisen*, vol. 32, nos 4 and 5, pp. 61–66, 80–82, 1933.

[32] F. Gebauer, "Berechnung der Eisenbetonbalken unter Berücksichtigung der Schwindspannungen der Eisen," *Beton und Eisen*, vol. 33, no. 9, pp. 137–143, 1934.

[33] O. Baumann, "Die Knickung der Eisenbeton-Säulen. Eidg," *Material-Prüfungsanstalt an der E. T. H. in Zürich, Bericht*, vol. 89, p. 56, 1934.

[34] E. Bittner, "Zur Klärung der n-Frage bei Eisenbetonbalken," *Beton und Eisen*, vol. 34, no. 14, pp. 226–228, 1935.

[35] E. Bittner, "Die Berechnung der Eisenbetonbalken nach Zustand III," *Beton und Eisen*, vol. 35, no. 9, pp. 143–146, 1936.

[36] A. Brandtzæg, "Der Bruchspannungszustand und der Sicherheitsgrad von rechteckigen Eisenbetonquerschnitten unter Biegung oder aussermittigem Druck," *Norges Tekniske Høiskole, Avhandlinger til 25-årsjubileet*, pp. 667–764, 1935.

[37] F.v. Emperger, "Der Beiwert *n*," *Beton und Eisen,* vol. 35, no. 19, pp. 324–332, 1936.
[38] R. Saliger, "Bruchzustand und Sicherheit im Eisenbetonbalken," *Beton und Eisen,* vol. 35, no. 19 and 20, pp. 317–320 and 339–346, 1936.
[39] U. Specifications, *OST 90003: Standards and Technical Conditions for the Design of Reinforced Concrete Structures,* Moscow: USSR Specifications, 1939.
[40] K.A. Hruban, "Plastic Theory in the New Czechoslovakian Regulations," *Concrete and Constructional Engineering,* vol. 44, no. 12, pp. 375–377, 1949.
[41] K.C. Cox, "Tests of Reinforced Concrete Beams with Recommendations for Attaining Balanced Design," *ACI Journal,* vol. 38, pp. 65–80, 1941.
[42] V.P. Jensen, *Ultimate Strength of Reinforced Concrete Beams as Related to the Plasticity Ratio of Concrete*, vol. 345, Champaign, IL: University of Illinois, Engineering Experiment Station Bulletin, 1943, p. 60.
[43] V.P. Jensen, "The Plasticity Ratio of Concrete and its Effect on the Ultimate Strength of Beams," *ACI Journal,* vol. 39, pp. 565–582, 1943.
[44] J. Melan, "Die Biegebruchspannungen des Eisenbetonbalkens," *Beton und Eisen,* vol. 35, no. 19, pp. 315–317, 1936.
[45] Portland Cement Association, "Ultimate Design of Reinforced Concrete," *Modern Developments in Reinforced Concrete,* vol. 11, p. 19, 1944.
[46] R. Chambaud, "Étude Expérimentale de la Flexion dans les Pièces en Béton Armé," *Annales de l'Institut Technique du Bâtiment et des Travaux Publics,* vol. 61, no. 4, p. 36, 1949.
[47] R. Chambaud, "Théorie Élastoplastique de la Flexion dans les Poutres en Béton Armé," *Annales de l'Institut Technique du Bâtiment et des Travaux Publics,* vol. 101, no. 10, 1949.
[48] F. Richart, A. Brandtzæg, and R. Brown, *A Study of the Failure of Concrete under Combined Compressive Stresses*, vol. 185, Champaign, IL: University of Illinois at Urbana Champaign, 1928, p. 104.
[49] M. Sargin, *Stress-Strain Relationships for Concrete and the Analysis of Structural Concrete Sections*, Solid Mechanics Division, University of Waterloo, 1971.
[50] S. Popovics, "A Numerical Approach to the Complete Stress-Strain Curve of Concrete," *Cement and Concrete Research,* vol. 3, no. 5, pp. 583–599, 1973.
[51] S. Popovics, "Mechanical Behavior of Materials," in *Proceedings of the International Conference on Mechanical Behavior of Materials*, 1972.
[52] P. Wang, S. Shah, and A. Naaman, "Stress-Strain Curves of Normal and Lightweight Concrete in Compression," *ACI Journal,* vol. 75, no. 11, pp. 603–611, 1978.
[53] D.J. Carreira and K. H. Chu, "Stress-Strain Relationship for Plane Concrete in Compression," *ACI Journal,* vol. 82, no. 6, pp. 797–804, 1985.
[54] J.B. Mander, M.J.N. Priestley, and R. Park, "Theoretical Stress-Strain Model for Confined Concrete," *Journal of Structural Engineering (ASCE),* vol. 114, no. 8, pp. 1804–1826, 1988.
[55] M.A. Anis, B.J. Farid, and A.I. Al-Janabi, "Stress-Strain Relationship for Concrete in Compression Made of Local Materials," *JKAU: Engineering Sciences,* vol. 2, pp. 183–194, 1990.
[56] I. Iskhakov, *Concrete Structures 1* (Selected Topics and Exercises), Ariel: Academic College of Judea and Samaria, 2002.
[57] G. Minafò and M. Papia, "Concrete Softening Effects on the Axial Capacity of RC Jacketed Circular Columns," *Engineering Structures,* vol. 128, pp. 215–224, 2016.

[58] I. Iskhakov and Y. Ribakov, *Ultimate Equilibrium of RC Structures Using Mini-Max Principle*, New York, NY: Nova Science Publishers, Inc., 2014.

[59] I. Iskhakov, "Quasi-Isotropic Ideally Elastic-Plastic Model for Calculation of RC Elements without Empirical Coefficients," *Structural Engineering, Mechanics, and Computation,* vol. 1, pp. 367–374, 2001.

[60] I. Iskhakov and Y. Ribakov, "Structural Phenomenon of Cement-Based Composite Elements in Ultimate Limit State," *Advances in Materials Science and Engineering,* vol. 2016, 2016.

[61] A.A. Gvozdev, "Obtaining the Destructive Loads Value for Statically Indetermined Systems under Plastic Deformations," in *Proceedings of the Conference on Plastic Deformations*, Moscow, 1938.

[62] I. Iskhakov, *Plastic Energy Dissipation, Ductility, and Load-Carrying Capacity of RC Elements*, Tel-Aviv: ALFA Publishers, 2012.

[63] I. Iskhakov, Y. Ribakov, K. Holschemacher, and T. Mueller, "Experimental Investigation of Full-Scale Two-Layer Reinforced Concrete Beams," *Mechanics of Advanced Materials and Structures,* vol. 21, no. 4, pp. 273–283, 2014.

[64] I. Iskhakov and Y. Ribakov, "Selecting the Properties of a Base Isolation System Based on Impulse Testing of a Three-Story Structural Part," *European Earthquake Engineering,* vol. 1, pp. 38–42, 2005.

[65] I. Iskhakov and Y. Ribakov, "Experimental and Numerical Investigation of a Full-Scale Multi-storey RC Building under Dynamic Loading," *The Structural Design of Tall and Special Buildings,* vol. 14, no. 4, pp. 299–313, 2005.

[66] I. Iskhakov and Y. Ribakov, "A New Concrete Creep Model with Minimum Empirical Coefficients," *Romanian Journal of Materials,* vol. 51, no. 4, pp. 528–535, 2021.

[67] The Standards Institution of Israel, *SI 26 Part 5: Methods of Testing Concrete: Properties of Hardened Concrete Other than Strength,* Tel Aviv: The Standards Institution of Israel, 1995.

[68] ASTM International, *ASTM C39/C39M-20: Standard Test Method for Compressive Strength of Cylindrical Concrete Specimens,* West Conshohocken, PA: ASTM International, 2020.

[69] G. Muciaccia, G. Rosati, and G.D. Luzio, "Compressive Failure and Size Effect in Plain Concrete Cylindrical Specimens," *Construction and Building Materials,* vol. 137, pp. 185–194, 2017.

[70] X. Shi, P. Park, Y. Rew, K. Huang, and C. Sim, "Constitutive Behaviors of Steel Fiber Reinforced Concrete Under Uniaxial Compression and Tension," *Construction and Building Materials,* vol. 233, p. 15, 2020.

[71] M. Roddenberry, R. Kampmann, M.H. Ansley, N. Bouchard, and W.V. Ping, "Failure Behavior of Concrete Cylinders under Different End Conditions," *ACI Materials Journal,* vol. 108, pp. 79–87, 2011.

[72] K. Holschemacher, I. Iskhakov, and Y. Ribakov, "Laboratory Tests of Two-Layer Beams Consisting of Normal and Fibered High Strength Concrete: Ductility and Technological Aspects," *Mechanics of Advanced Materials and Structures,* vol. 19, pp. 513–522, 2012.

[73] I. Iskhakov and Y. Ribakov, "Transverse Deformations and Structural Phenomenon as Indicators of Steel Fibred High-Strength Concrete Non linear Behavior," *Advances in Materials Science and Engineering,* vol. 2019, 2019.

[74] I. Iskhakov, I. Frolov, and Y. Ribakov, "Experimental Investigation of Concrete Transverse Deformations at Relatively High Loading Rates for Interpretation of High Strength Concrete Behavior," *Applied Sciences,* vol. 11, no. 8460, p. 16, 2021.

[75] V. Revilla-Cuesta, M. Skaf, A. Santamaría, V. Ortega-López, and J.M. Manso, "Assessment of Longitudinal and Transversal Plastic Behavior of Recycled Aggregate Self-Compacting Concrete: A Two-Way Study," *Construction and Building Materials,* vol. 292, 2021.

[76] O. Leibovich, D.Z. Yankelevsky, and A.N. Dancygier, "Predicting the Circumferential Strains on a Cylindrical Concrete Spec-imen During a Pull-Out Test," *Structural Concrete,* vol. 22, pp. 3026–3041, 2021.

[77] I. Iskhakov, Y. Ribakov, K. Holschemacher, and S. Kaeseberg, "Elastic-Plastic Stage of Transverse Deformations in the Compressed Zone of Real Reinforced Concrete Beams," *Applied Sciences,* vol. 13, 2023.

[78] M. Zhou, J. Liao, L. An, W. Deng, M.F. Hassanein, and Z. Yu, "Analysis of Stress-Induced Cracks in Concrete and Mortar under Cyclic Uniaxial Compression," *Construction and Building Materials,* vol. 187, pp. 652–664, 2018.

[79] British Standards Institution, *EN 206: Concrete – Specification, Performance, Production, and Conformity,* London: British Standards Institution, 2013.

[80] J.G.M. Van Mier, S.P. Shah, M. Arnaud, J.P. Balayssac, A. Bascoul, S. Choi, D. Dasenbrock, G. Ferrara, C. French, M. E. Gobbi, et al., "Strain-Softening of Concrete in Uniaxial Compression – Report of the Round-Robin Test carried out by RILEM TC 148-SSC," *Materials and Structures,* vol. 30, pp. 195–209, 1997.

[81] CEN, *EN 197: Part 1: Cement. Composition, Specifications, and Conformity Criteria for Common Cements,* Brussels: CEN, 2011.

[82] The Standards Institution of Israel, *SI 3: Mineral Aggregates from Natural Sources,* Tel Aviv: The Standards Institution of Israel, 1998.

[83] E.I. Baranovskiy, V.V. Babicky, A.I. Bondarovich, and P.V. Ryabchikov, *Methodical Instructions for Laboratory Works on the Discipline: Factory Production Technology of Concrete and Reinforced Concrete Structures*, Minsk: BNTU, 2014.

[84] ASTM International, *ASTM C31/C31M-19: Standard Practice for Making and Curing Concrete Test Specimens in the Field,* West Conshohocken, PA: ASTM International, 2019.

[85] British Standards Institution, *EN 12390: Part 2: Testing Hardened Concrete. Making and Curing Specimens for Strength Tests,* London: British Standards Institution, 2019.

[86] ASTM International, *ASTM C192/C192M-19: Standard Practice for Making and Curing Concrete Test Specimens in the Laboratory,* West Conshohocken, PA: ASTM International, 2019.

[87] The Standards Institution of Israel, *SI 1865 Part 2: Concrete Code: General Principles Paving Test Methods: Tests of Aggregates,* Tel Aviv: The Standards Institution of Israel, 1998.

[88] ASTM International, *ASTM C136/C136M-19: Standard Test Method for Sieve Analysis of Fine and Coarse Aggregates,* West Conshohocken, PA: ASTM International, 2019.

[89] British Standards Institution, *EN 196: Part 3: Methods of Testing Cement: Determination of Setting Times and Soundness,* London: British Standards Institution, 2017.

[90] British Standards Institution, *EN 1097: Part 3: Tests for Mechanical and Physical Properties of Aggregates: Determination of Loose Bulk Density and Voids,* London: British Standards Institution, 1998.
[91] British Standards Institution, *EN 1097: Part 6: Tests for Mechanical and Physical Properties of Aggregates — Determination of Particle Density and Water Absorption,* London: British Standards Institution, 2017.
[92] British Standards Institution, *EN 12350: Part 2: Testing Fresh Concrete. Slump Test,* London: British Standards Institution, 2019.
[93] British Standards Institution, *EN 12350: Part 6: Testing Fresh Concrete. Density,* London: British Standards Institution, 2019.
[94] A. Pisanty, *Reinforced Concrete* Principles, Haifa: Faculty of Civil Engineering, Technion, 2007.
[95] British Standards Institution, *EN 12390: Part 4: Testing Hardened Concrete Compressive Strength. Specification for Testing Machines,* London: British Standards Institution, 2019.
[96] RILEM TC 148-SSC, "Test Method for Measurement of the Strain-Softening Behaviour of Concrete under Uniaxial Compression," *Materials and Structures,* vol. 33, pp. 347–351, 2000.
[97] International Standard, *ISO 1920: Part 10: Testing Concrete. Determination of Static Modulus of Elasticity in Compression,* Geneva: International Standard, 2010.
[98] M.A. Mansur, T.H. Wee, and M.S. Chin, "Derivation of the Complete Stress-Strain Curves for Concrete in Compression," *Magazine of Concrete Research,* vol. 47, no. 173, pp. 285–290, 1995.
[99] J. Hadamard, *Leçons sur la Propagation des Ondes et les Équations de l'Hydrodynamique* (Chapter VI), Paris: Hermann et Cie, 1903.
[100] P.G. Bergan, "Record of the Discussion on Numerical Modeling," in *IUTAM W. Prager Symposium*, Evanston, Illinois, 1983.
[101] G.A. Hegemier and H. E. Read, "Some Comments on Strain-Softening," in *DARPA-NSF Workshop*, Evanston, Illinois, 1983.
[102] I. Sandler and J.P. Wright, "Summary of Strain-Softening," in *DARPA-NSF Workshop*, Evanston, Illinois, 1983.
[103] J.A. Hudson, E.T. Brown, and C. Fairhurst, "Shape of the Complete Stress-Strain Curve for Rock," in *13th Symposium on Rock Mechanics*, Urbana, Illinois, 1971.
[104] E. Ferretti, "Experimental Procedure for Verifying Strain-Softening in Concrete," *International Journal of Fracture,* vol. 126, pp. 27–34, 2004.
[105] E. Ferretti, "A Discussion of Strain-Softening in Concrete," *International Journal of Fracture,* vol. 126, pp. L3–L10, 2004.
[106] I. Iskhakov and Y. Ribakov, "Collapse Analysis of Real RC Spatial Structures Using Known Failure Schemes of Ferro-Cement Shell Models," *The Structural Design of Tall and Special Buildings,* vol. 23, no. 4, pp. 272–284, 2014.
[107] S. Shah, U. Gokoz, and F. Ansari, "An Experimental Technique for Obtaining Complete Stress-Strain Curves for High Strength Concrete," *Cement, Concrete, and Aggregates,* vol. 3, pp. 21–27, 1981.
[108] M.D. Kotsovos, "Effect of Testing Techniques on the Post-Ultimate Behaviour of Concrete in Compression," *Matériaux et Construction,* vol. 16, pp. 3–12, 1983.
[109] A. Hillerborg, *Stability Problems in Fracture Mechanics Testing. Fracture of Concrete and Rock: Recent Developments*, S. P. S. S. E. B. B. (Please check if

initials are correct, appear too lengthy) Shah, Ed., Elsevier Applied Science, 1989, pp. 369–378.

[110] V.M. Bondarenko and D.G. Suvorkin, *Reinforced Concrete and Stone Structures*, Moscow: Vysshaya Shkola, 1987.

[111] NIIZHB, *BR 52-101-2003: Non-Pre-Stressed Concrete and Reinforced Concrete Structures,* Moscow: NIIZHB, 2004.

[112] I. Iskhakov and Y. Ribakov, "Structural Phenomenon Based Theoretical Model of Concrete Tensile Behavior at Different Stress-Strain Conditions," *Journal of Building Engineering,* vol. 33, no. 101594, 2021.

[113] I. Iskhakov, I. Frolov, and Y. Ribakov, "Experimental Verification of Theoretical Stress-Strain Model for Compressed Concrete Considering Post-Peak Stage," *Materials,* vol. 15, no. 6064, 2022.

Appendix. List of Symbols

The symbols used in this book are defined when they are first introduced. The most common symbols are listed below for convenient reference.

A_c	Area of concrete section
A_i	Cumulative residue of each aggregate fraction on sieve
A_s	Area of tensile reinforcement section
A'_s	Area of compression reinforcement section
C	Cement consumption for concrete mix proportioning
E_c	Tangent modulus of elasticity of normal weight concrete
E_{sec}	Secant modulus of elasticity of concrete
E_s	Modulus of elasticity of reinforcing steel
FM	Fineness modulus of fine aggregate
G	Coarse aggregate consumption for concrete mix proportioning
K_{scc}	Standard cement consistency
M	Bending moment
M_{cd}	Part of internal bending moment transferred to compressed concrete
M_d	Design value of applied internal bending moment
M_{el}	Maximal elastic moment in bended beam
$M_{\max}$	Maximum bending moment
M_{pl}	Maximal plastic moment in bended beam
M_{sd}	Part of internal bending moment transferred to tension reinforcement
ΔM	Part of internal bending moment transferred to compressed reinforcement
P	Load bearing capacity
P_u^-	Lower bound of load-bearing capacity

P_u^+	Upper bound of load-bearing capacity
RW	Retarding and water-reducing admixture consumption for concrete mix proportioning
S	Fine aggregate consumption for concrete mix proportioning
S_0	First moment of cross-sectional area
$S_{c,\max}$	First moment of compressed cross-sectional area
T	Vibration period
U_{el}	Concrete elastic potential
U_{pl}	Concrete plastic potential
V_{Rd}	Design value of applied shear force
V_{Rdc}	Part of design shear force value transferred to concrete
V_{Rds}	Part of design shear force value transferred to reinforcement
$\frac{W}{C}$	Water-cement ratio
W	Final water demand for concrete mix proportioning
W_T	Preliminary water demand for concrete mix proportioning
WR	Water-reducing and retarding admixture consumption for concrete mix proportioning
a_i	Part of material on sieve, relative to initial weight
b	Overall width of cross-section
d	Effective depth of cross-section
d'_s	Concrete cover depth for compression reinforcement
e	Eccentricity
f_c	Compressive strength of concrete
f'_c	Compressive strength of concrete in flexure
f_{cem}	Cement class
f_{ck}	Characteristic compressive cylinder strength of concrete at 28 days
f_{ck}^{prism}	Characteristic compressive prism strength of concrete at 28 days
f_{ct}	Axial tensile strength of concrete
f_s	Tensile stress in reinforcement
f_y	Yield strength of longitudinal reinforcement
k	Coefficient depending on aggregate quality

k_1	Coefficient depending on aggregate quality
Δl_{init}	Initial longitudinal shortening
m	Initial aggregate weight
m_i	Mass of material retained on sieve
n	Modular ratio
q	Uniformly distributed load
r	Parameter defining the failure mechanism scheme
x	Neutral axis depth
y_{extr}	Distance from neutral axis to fiber with maximum stress
α	Coefficient of coarse aggregate grains expansion by cement-sand mortar
δ	Deviation
ε_c	Instantaneous values of strain in concrete
ε_{c1}	Compressive strain in concrete at peak stress f_c
$\varepsilon_{c\,el}$	Maximum value of concrete longitudinal elastic strain
$\varepsilon_{c\,el\,ul}$	Ultimate elastic strain determining concrete elastic potential
$\varepsilon_{c\,ul}$	Ultimate compressive strain in concrete
$\varepsilon_{ct\,ul}$	Ultimate elastic transverse deformation in concrete
ε_{ctm}	Maximal tensile strain in concrete
$\varepsilon_{long\,init}$	Initial longitudinal strain
$\varepsilon_{shear\,res}$	Residual shear deformation after unloading
$\varepsilon_{shear\,ul}$	Ultimate shear deformation
ε_{trans}	Instantaneous values of transverse deformation
ε_{trans1}	Transverse strain in concrete at peak stress f_c
$\varepsilon_{trans\,res}$	Residual transverse deformation after unloading
$\varepsilon_{trans\,ul}$	Ultimate transverse deformation
η	Plasticity ratio
ξ	Damping ratio
ψ	Internal forces filed
ρ_c	Fresh concrete density in compacted state
ρ_g	Coarse aggregate particle density
ρ_{gb}	Loose (oven-dry) bulk density of coarse aggregate
ρ_s	Fine aggregate particle density

ρ_{st}	Ratio of effective longitudinal reinforcement area to gross concrete area
σ_c	Instantaneous values of stress
$\sigma_{c\,el}$	Ultimate elastic stress
$\sigma_{c\,ul}$	Ultimate stress corresponding to boundary of stress-strain diagram descending branch
σ_{init}	Initial stress in loading process
$\sigma(h/2)$	Stress in section top fiber
Δ	Span displacement
Δ_{res}	Residual span displacement after unloading
Δ_{ul}	Ultimate span displacement
ν	Poisson coefficient
ν_0	Minimal experimentally obtained value of Poisson coefficient for concrete without steel fibers
ν_0	Maximal experimentally obtained value of Poisson coefficient for concrete without steel fibers
ν_{30}	Minimal experimentally obtained value of Poisson coefficient for concrete with steel fiber content of 30 kg/m^3
ν_{30}	Maximal experimentally obtained value of Poisson coefficient for concrete with steel fiber content of 30 kg/m^3
ω	Relationship $\frac{x}{d}$ for bended rectangular section

Index

For Product Safety Concerns and Information please contact our EU representative GPSR@taylorandfrancis.com
Taylor & Francis Verlag GmbH, Kaufingerstraße 24, 80331 München, Germany

www.ingramcontent.com/pod-product-compliance
Ingram Content Group UK Ltd.
Pitfield, Milton Keynes, MK11 3LW, UK
UKHW022157190926
475702UK00003B/82
* 9 7 8 1 0 3 2 7 3 8 9 1 8 *